AF478617

Knowledge Loves Company

Knowledge Loves Company

Successful Models of Cooperation between Universities and Companies in Europe

Edited by

Hans-Joachim Gögl and Clemens Schedler
Commissioned by Landschaft des Wissens

Photography by Claudio Alessandri

First published in German 2005

English edition published 2009 by
PALGRAVE MACMILLAN

Palgrave Macmillan in the UK is an imprint of Macmillan Publishers Limited,
registered in England, company number 785998, of Houndmills, Basingstoke,
Hampshire RG21 6XS.

Palgrave Macmillan in the US is a division of St Martin's Press LLC,
175 Fifth Avenue, New York, NY 10010.

Palgrave Macmillan is the global academic imprint of the above companies
and has companies and representatives throughout the world.

Palgrave® and Macmillan® are registered trademarks in the United States,
the United Kingdom, Europe and other countries.

ISBN-13: 978–0–230–57586–8   hardback
ISBN-10: 0–230–57586–2   hardback

This book is printed on paper suitable for recycling and made from fully
managed and sustained forest sources. Logging, pulping and manufacturing
processes are expected to conform to the environmental regulations of the
country of origin.

A catalogue record for this book is available from the British Library.

A catalog record for this book is available from the Library of Congress.

10   9   8   7   6   5   4   3   2   1
18   17   16   15   14   13   12   11   10   09

Printed and bound in Great Britain by
CPI Antony Rowe, Chippenham and Eastbourne

Contents

Preface

It is a tautology to state that living things want to live; by contrast, it is an ideology to claim that the primary drive of living systems is to propagate themselves and fight against each other.

A look at evolution and the developmental stages of living systems reveals that cooperation was the decisive prerequisite for the emergence of life and it has remained a phenomenon that accompanies life in all its variations... Interaction and resonance characterize all significant biological structures and processes.

(Joachim Bauer)[1]

The series *Landschaft des Wissens* (Landscape of Knowledge) aims at locating and describing European models of cooperation with a high potential for economic development. It is no coincidence that this idea was born in the periphery, where there is a pronounced sensibility for the need and the chance to make up for existing deficiencies and perhaps also where the successful interaction of talents is perceived and appreciated more strongly as the scarce resource that it is. The first volume of this series focuses on *Big Strategies for Small Business*[2] in seven reportage-style portraits of innovative cooperation projects in regions throughout Europe, from Friuli in Italy to Petäjävesi in Finland.

The association that publishes this series is located in Carinthia, a typical European region where many different languages and cultures converge. It is also home to a young technology project that over the past few years has become one of the largest of its kind in Austria. Selected international IT companies were recruited to settle on the campus of a university that is small but also highly specialized in specific fields. This proximity fosters close collaboration on research and development projects.[3] This site of interaction, which was consciously initiated as a

---

[1] Excerpted from the book *Prinzip Menschlichkeit. Warum wir von Natur aus kooperieren* [The Human Principle: Why We Are Naturally Cooperative] by Joachim Bauer, neurobiologist, Hoffmann und Campe, Frankfurt/M 2006.

[2] *Big Strategies for Small Business: Exceptional Projects in Europe*, Palgrave Macmillan, Basingstoke 2009.

[3] Lakeside Science & Technology Park, "A New Architecture of Cooperation," pages 127–154.

driving impulse for regional development, has inspired us to look at and study the vast range of cooperation models in Europe – with their varying aims, forms, and locations.

Where there is fertile exchange, there is growth. Companies, students, and instructors profit, resulting in new knowledge and in many cases jobs: knowledge spawns companies. This is something that can be seen increasingly in small and medium-sized enterprises (SMEs), whose considerable contribution to the prosperity of rural areas and communities in Europe is sometimes underestimated. Our focus was on the specific collaboration opportunities open to small businesses which for organizational or financial reasons have not been aggressive enough when considering working with universities. In each interview we offer encouragement, practical advice, and examples.

From Dr. No to Dr. Yes

The classic form of cooperation is usually the joint research project, which gives universities urgently needed third-party funds and provides companies with immediate access to potential employees. This clear overlap of interests has meanwhile spawned countless projects and a gradually emerging reflective expertise not only at universities but also at large companies in particular. These activities – currently being practiced by a global scientific community strongly influenced by the English-speaking world – have in recent years helped to open the ivory towers of the European mainland, at least enough to let in a bit of fresh air. The European system is one that has traditionally tended to value self-sufficiency. This hermetic stance can be seen in the media in such powerful stereotypical images as the absent-minded professor, who you can't even let cross the street alone, or James Bond's brilliant, megalomaniacal adversaries who want to rule the world. The ability to work together is not necessarily a dominant quality of either of these characters. And, of course, those who seem to fit this one-dimensional mold are partly to blame for the continued use of these stereotypes.

The ties that have been developing between universities and companies and growing stronger over the years also harbor ambivalences, which the interviews in this book have repeatedly sought to examine in a critical manner. Open research, the commitment to a subject that is not commercially viable, pleasure in knowledge of and for itself – all these are part of the basic cultural values of our society. Today more than ever it seems as if we have to demand, discuss, and actively defend spaces that are free from economic paradigms, from the pervasion of the shortage and survival principle that tells us that something is only worth anything if it can turn a profit.

> Cooperation partners bring the worldviews of their systems along with them into every project they work on together. Each of them has clear objectives, but they are different. Thus, cooperation is more likely to engender a conflict situation than one of agreement; this is the assumption we must make.
>
> (Peter Heintel *on page 10 of this book*)

One of the main focuses of this volume is to examine and elucidate different aspects of a few extraordinary examples of research collaboration: from ingenious technology transfer strategies in Germany or Great Britain to working together on a basic research project about business management in Switzerland to an amazing Belgian model for cooperation between competitors. Parallel to this, we looked for further reasons and approaches for pursuing company–university win-win models and uncovered exemplary projects in the fields of the cultural sciences, development aid, and regional development. We also took a look at a new EU member state where a couple of decades ago the private sector didn't even exist.

Of course, this volume of 11 interviews with 25 academics and business leaders is a subjective selection and does not claim to be exhaustive in either subject matter or form. It is a small but varied section chosen from the mosaic of the more than 100 projects researched for this book. And they, in turn, are only a fraction of the thousands being practiced daily at Europe's universities, technical colleges, and other research and training centers.

In our research, we initially focused on the exemplary, strategic quality of the cooperation models: where can we find especially innovative structures of cooperation from which other companies or universities might learn? We searched for processes, methods, and organizational instruments at universities and within academic–company projects.

But as we delved deeper into the projects and got to know the people engaged with each other in daily exchange and communication, another much more elusive aspect began to emerge: the process of cooperation between different ways of thinking or a glimpse into the intermediate spaces of successful cooperation.

In all of our interviews there were certain concepts, actually values, that seemed to pop up again and again, things we never expected (as a negative result, perhaps, of the above-mentioned professor/scientist cliché) as we were preparing our interviews with an expert on nanotechnology, an executive of one of the global players in the field of aerospace propulsion technology, or a bioengineer who is also on the way to making the scourge of high blood pressure a relic of the past. For example: nonfunctionalized time spent together, spaces for encounter (physical!), fights – what can be gained and the skills required to deal with them, agreements jointly worked out verbally and then formulated in writing, courage, entrepreneurship, understanding the different work rhythms and paces of commercial and academic processes, long-term, personal relationships.

> I've never done an investment deal…which did not first involve a significant amount of time over a beer.
>
> (Timothy Barnes *on page 123 of this book*)

On the one hand we found these so-called soft skills and our interview partners' insistence on these personal aspects in joint projects confusing. After all, we had set out to find empirical recipes for cooperation that could be implemented broadly, regardless of staff or location. On the other hand these factors had also popped up in the first volume about strategies for small business, emerging as the vital, complex but also decisive bonds holding together the important but ultimately controllable organizational structures – as the managers of SMEs had told us then and entrepreneurs and academics were telling us now.

That is how the text and visual design editors chose to focus on soft skill issues in cooperation projects as the inspiring element of this volume. It was decided that the form for describing the selected cooperation projects would be the interview rather than the essay or report. The interview with the academics and their commercial partners seated around the same table conveys their experiences in working together and the essence of these partnerships in a vibrant and in our estimation also fitting way. Fitting because despite the useful insight on cooperation outlined here, we don't want to suggest one could go about tackling the challenge with a carefully devised quality management system checklist.

The relation between the insights and the people to whom they are a part of everyday practice, the diversity with which each person expresses him- or herself, our authors' positions as question-posers rather than answer-, summary-, or conclusion-givers provides us with a form that fits well with the theme of this book, though we are also fully aware that by virtue of the diversity of the personalities compiled here, the repetition, reinforcements, and arising patterns, the metalevel of reflection is bound to take place in the reader rather than emerge as an independent text genre.

This subjective mood is recorded visually in the positivist perceptual tradition of photographs that show the locations of cooperation, the hands and faces of the people involved, and a short "film" sequence of each interview.

At the end of each chapter we have provided the contact information of our interview partners – in the spirit of this book: cooperation, exchange, and resonance.

Hans-Joachim Gögl & Clemens Theobert Schedler

Acknowledgments

The jury of the competition "Schönste Bücher Österreichs 2005" [Austria's Best Designed Books 2005] and the Hauptverband des Österreichischen Buchhandels for awarding our first volume with the Austrian National Prize.

The Stiftung Buchkunst in Frankfurt am Main/Leipzig for the competition "Golden Letter" [World's Best Designed Books 2006] and the jury for voting to award volume 01 of this book series with the silver medal.

Silence for giving us the endless space to talk this book and everything about it into existence.

Are you serious? If so, you might as well mention the universe and the car manufacturer Saab too.

Silence and the universe, aren't they the same thing?

Potentials, Contradictions, and Game Rules:
On Models of Cooperation Between Universities and Companies

Interview by *Hans-Joachim Gögl*

Austria

"If it is true that in complex decision-making situations the starting position is marked by contradictions and conflicts, then it would seem that the most important skill to be mastered in cooperation projects would be that of contradiction management."

Which different cultures and ways of thinking converge when universities and companies work together?

Is there such a thing as rules for successful interdisciplinary collaboration? And in what ways are the growing influence that academia and industry have on each other starting to affect the realities of both systems?

Questions as a prologue to the cooperation models presented here, directed at a philosopher and management consultant, cultural scientist and pioneer for transdisciplinary research projects that span the diverse ways academics and businesspeople see themselves. A critical expert with many years of practical experience in both worlds discussed in this book.

University Professor Dr. Peter Heintel teaches at the Institute of Philosophy and Group Dynamics and the Faculty for Interdisciplinary Research and Further Training (IFF) at the University of Klagenfurt. He deals in theory and practice with interdisciplinary cooperation and project and conflict management. In addition to his university research and teaching activity, he has also served institutions and companies for many years as a consultant for organizational issues.

Professor Heintel, let's say you were a businessman running a small, successful software company with 50 employees when a good fairy appeared and said you could ask the university for three things. What would you wish for?

In this field in particular we are currently faced with a pan-European problem because the number of students going into computer science is decreasing noticeably. One strategy for reversing this trend would be for the universities to get in touch with business reality and take supportive action with the help of their students, especially in research and development projects. I also think industry executives should be incorporated more often into universities as instructors because the computer science curriculum is so specialized it fails to prepare students for the integrative positions companies need filled. So my first wish would be for more cooperation at the training level; my second, for an increased focus on practical application at colleges and universities. And as long as

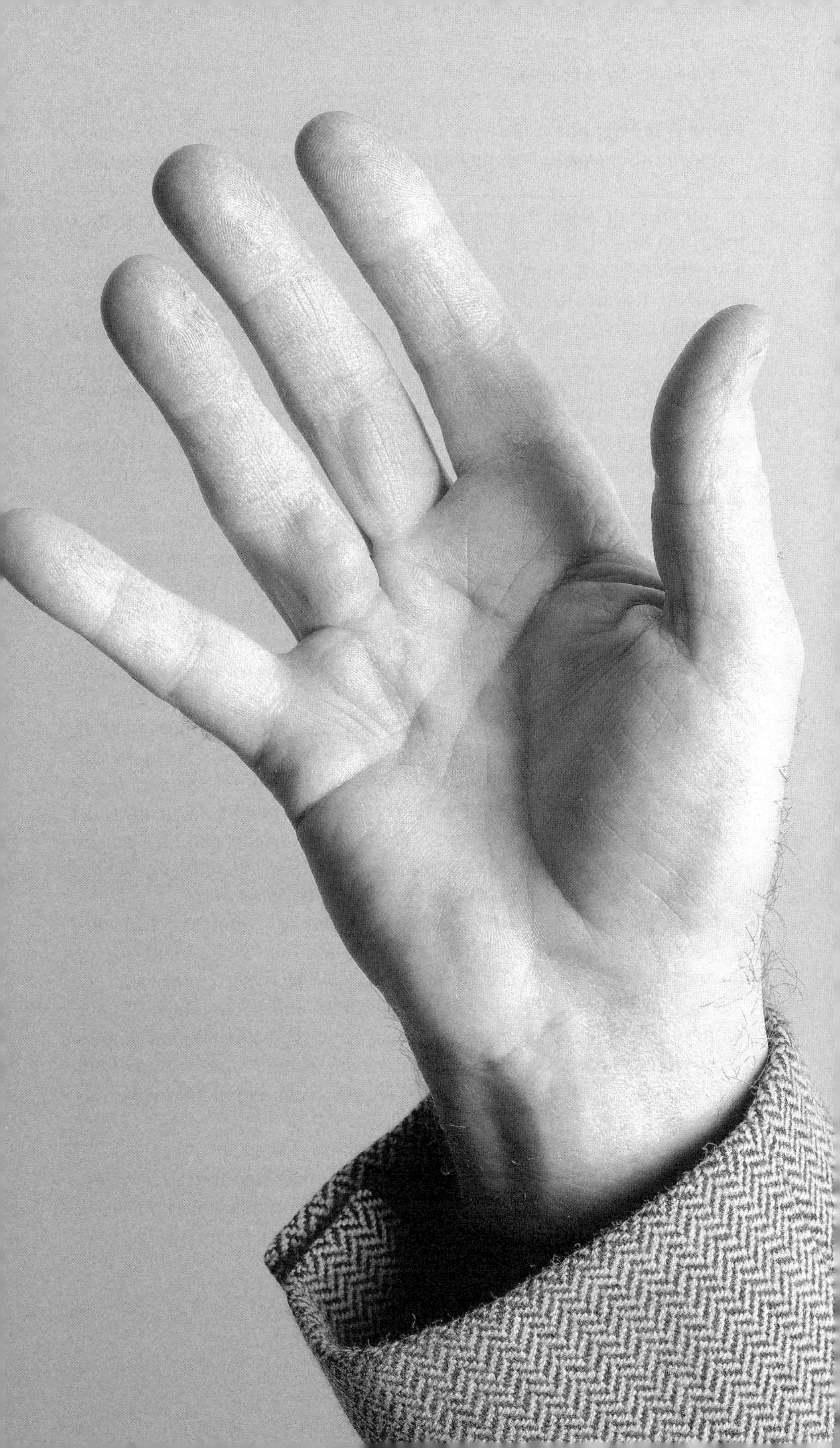

I have a fairy granting me wishes, I might as well ask for an experimental development-oriented company founded jointly by students and practitioners.

In one of your research projects you examine the reasons for the enormous decline of students in technical fields. In China or India there's a veritable boom in these disciplines; in Europe, by contrast, there is a shift toward human and life sciences, e.g. psychology. They are being taken by storm despite the fact that job prospects continue to be bleak.

Interviews with young people reveal a tragicomic image of the computer science major: a four-eyed, somewhat overweight guy who might not be able to find a girlfriend, but can hack into the Pentagon with one hand while clutching a Big Mac in the other. What can universities do about these images and the general climate of our times?

What is most important is getting rid of segmentation. That doesn't apply to computer science only, but it's a good example. Now, even primary schools are offering something like computer classes, you have computer studies at technical colleges and universities, and then you've got parents who have no idea about any of this because they've never been confronted with the subject, not to mention the grandparents, who often play a decisive role in a child's career choice.

We need intermediary platforms that make sure that these groups have the opportunity to discuss the results of the surveys and studies. This generates a shared view of the subject, the problems, and prejudices. The next step is designing an implementation plan – the practical planning of the measures arrived at jointly. People don't talk to each other! Students and their parents feel insecure about future career prospects, and there is hardly any contact between these target groups and the people in the business world.

Only a collective view allows us to work out problems in a way that is decisive and also new *because usually the results of motivational research are presented once, then made available to the participants, and that's it. Our approach to academics here in Klagenfurt may be unusual, but organizations determine their own content, solution options only arise through joint communicative processing. It is not as if one translates already existing content into structures, that's a technological notion of implementation that doesn't work in a social context.*

That means new solutions require new or adapted organizational forms.

Yes, powerful solutions require an ongoing dialogue between all the people involved in the problem and solution; that itself is a new organizational form.

Merely recognizing the challenge and attempting to apply it in the existing structure is usually doomed to fail.

What does this mean in concrete terms when university meets business? Here you still find the same old traditional cultures of organization, from the bureaucratic red tape of university administration to the sometimes military-inspired hierarchy of companies.

Universities have developed their own science-oriented way of thinking which utilizes a different time frame. This is an aspect quite often criticized by industry. Companies, on the other hand, tend to think in a very direct outcome-oriented way. They have to assert themselves in the overall economic system. When these two cultures come together, it often results in friction. This is why I am convinced that we need intermediaries in cooperation projects, organizational forms in which counseling is provided for communication between different and even contradictory ways of thinking. Here at the institute we see this as the job of applied science. Within companies an ongoing problem is the classic conflict between the production and sales departments, for example, or between the inside staff and field representatives. What we need, particularly in cooperation projects, is to make communication processes an institutionalized part of project management. In the future it will become a key area of expertise in companies as well as at cooperating universities.

Survival of the most cooperative?

In an economic context, the Darwinian notion of "survival of the fittest" is deeply ingrained in our society; we have phrases like "the big fish eat the small fish," or more recently "the fast eat the slow," which illustrate how ruthless the business world can be. Other works by Darwin – and recent brain research – emphasize our ability to work together. Interestingly, hardly anyone took note of these theories back then.

In your view as a philosopher and management consultant, is cooperation a kind of basic primal skill and motivation that we can always activate, or is our most primary instinct really to beat the competition?

Philosophers call it an aporia, a logical impasse. There is no either-or. Both are justified and right. That means we must try to find balanced solutions, and that can't always be achieved. There are people who say that there is too much cooperation, security, laziness – a phenomenon referred to by Konrad Lorenz as self-domestication – in the social welfare state today; so then we have the backlash: cut-throat competition or dog-eat-dog capitalism in globalized markets. Cooperation, equality, symmetry are things people want as long as they are not at the top. The individual, and this is something very European, wants to be special, wants to stand out. The question is how? In a business-dominated system, the individual distinguishes itself by what it has to show: property. That is why this is an especially protected commodity in our society.

We have to expect both: people want to be the same and they want to be different. We know since Aristotle that sooner or later an exclusively competition-oriented system will destroy itself. We see this today in the destruction of the environment, where we are pulling the foundation of our very existence out from under our feet.

But exclusively cooperative strategies are not compatible with our market system. That's why economists consider this motivation to be least probable. I don't support this quite as vehemently as they do, economic individualism is a social abstraction. Every contract is a form of cooperation. And without trust there can be no contract.

But isn't the subject of cooperation especially important in this day and age? The enormous specialization of knowledge alone renders technological progress impossible without the use of new, actively encouraged, and competent forms of cooperation?

That is definitely true, although Hegel also based his whole "Phenomenology of Spirit" on the alternation between the two opposing historical movements of cooperation and conflict. A wonderful book in this context.

Unlike in academia, there has always been a built-in bond that made cooperation in business obligatory: the jointly manufactured product, which might be seen as a fundamental affirmation of this strategy and the logic of business itself.

This forces people to work together, not like in the academic world, and this has its advantages as well as its price. We have created, anchored in our society's constitution, privileged spaces where problems can be contemplated freely, without reservations or bias. I think it would be a big mistake to sacrifice this, and it does happen sometimes when attempts are made to streamline academics. The price is also that this leads to tremendous specialization. We know more and more about less and less, and sometimes absolutely nothing about how things are related. In addition to this, the influence of the economy tends to individualize careers and incentive demands. How much have you published? Which international conventions have you given talks at? If these had been the criteria for judging Kant's "Critique of Pure Reason," he would have been kicked out of the university with no further ado.

These values prevent cooperation because in any halfway decent interdisciplinary research project no one knows by the end who thought of what at the right time. Of course there will always be this kind of solo specialist, and in some cases it is also legitimate, someone who distinguishes himself through what I call the infinitesimal exertion of power – the smaller you chop up the problem, the more control you have. But the trend is forging ahead in the direction of transdisciplinary research, i.e. the study of the research subjects themselves. We call this assisting the self-discovery process: incorporating the very subjects in question across disciplinary lines.

Science will assume more and more responsibility for fomenting communication between the sectors of our subdivided society. I think that a new role for academics in the future is that of the mediator – an area in which particularly the cultural sciences could exert their influence!

Seeking conflicts

This calls for a new way of looking at things and above all for new competencies in both the business and academic sides of the partnership. In an interview you once said, "People tend to think cooperation happens on its own, as long as people more or less get along. This is wrong."

Cooperation partners bring the worldviews of their systems along with them into every project they work on together. Each of them has clear objectives, but they are different. Thus, cooperation is more likely to engender a conflict situation than one of agreement; this is the assumption we must make. Not because people don't like each other, but because the perspectives and standpoints are so utterly different.

This usually triggers four ways of reacting, which we have acquired over the course of human history. First, act as if it were nothing: pseudo consensus harmony; second, fight: I assert myself because I'm right. That can lead to endless meetings and eventually to exhaustion; therefore third, resignation: accept fate. The fourth solution, a favorite among managers, make a quick decision: we need to get rid of the problem, who cares about the decisions I made yesterday? These four reaction patterns are very common, but they don't exactly encourage communication. So if it is true that in complex decision-making situations the starting position is marked by contradictions and conflicts, then it would seem that the most important skill to be mastered in cooperation projects would be that of contradiction management. From a historical perspective, it has been only very recently that we have learned how to deal with cooperation in a conscious way.

Before that, we tried to solve our conflicts with winners and their hegemonies …

… which has led to well-known disasters. Technologically, economically, and organizationally speaking, we have come a long way. In Europe we have achieved what mankind has always dreamed of. Terrific! But when it comes to

dealing with conflicts, we behave like cavemen, and this is where we need to invest energy and money, and by that I mean more than for mere technological progress. And in the interest of our social development, nothing says we have to make enormous strides in one field just because we have fallen behind in other areas. In this sense we need a second enlightenment to develop our capacity to work together.

That means at least a self-critical reflection of the four behavioral patterns you mentioned above ought to be a part of the basic knowledge possessed by every human being, and in the more narrow context of our topic: by everyone with a responsible role in R&D or who is involved in university collaboration projects.

Yes, and every interdisciplinary project should begin with an in-depth workshop that clearly states the different ways of thinking and conflicting goals and discusses how to deal with them.

What is most important in modern forms of cooperation is something I call "self-analysis." Today, people will often wrangle for position within a team, they don't understand that it is perfectly normal to start out with differences and contradictions, and that they aren't going to go away either! The question is how can we work together with this background? It's something one can think about and practice, and there are many models and instruments in the consulting business that companies employ daily. At universities, however, because the culture of cooperation described above doesn't exist to a significant degree, this happens far too infrequently.

On the subject of successful cooperation the spokeswoman of the Managing Board of Infineon Austria, Monika Kircher-Kohl, once said: "Every team needs someone who keeps his or her eye on reaching the goal, then you need those who make sure none of the players get left behind, two who are willing to take risks, one who gives the warnings, and finally someone who can market whatever the group comes up with – even if it's nothing sensational." I find this quote striking because she doesn't talk about the professional skills of the team members, nor about reflection and team-building structures, but about personal talents and mentalities. The other interviews in this volume also stress the importance of such so-called soft skills. What role do these factors play in successful cooperation?

The opposite of teamwork is hierarchy, which is a model that monopolizes the functions you mentioned. It doesn't just appropriate power but assigns responsibility in a classic way, which can also be interpreted in an ethical sense.

We are constantly discussing alternatives to assuming the entire responsibility with managers because this model simply doesn't work in today's complex environment. When it comes to guiding and mediating teams, the important consideration is that functions can change depending on the task and the occasion and the strengths each individual involved brings to the table. Hierarchies prevent precisely this and therefore aren't capable of making full use of their employees' valuable skills.

In the second season of *Star Trek* the crew was chosen based not only on professional expertise but also according to their personalities. From your perspective, would that be a group-dynamics-based, future-oriented model for team building?

Everyone is capable of more than you think. That means in certain situations they can activate skills they never knew they had. I have found that people are more universal and can only be typologically categorized to a certain degree. Besides, if someone knows that he or she was chosen because of a given qualification, he/she will try to prove, perhaps one-sidedly, that he/she is valuable to the group particularly in that area, and that can interfere.

University culture meets company culture

In addition to the challenge of integrating different disciplines and personalities, there is also another major difference: the discrepancy between the culture of the university biotope and that of the company. Aren't these different natures the most valuable potential for collaboration, especially because this brings together different perspectives?

We are all familiar with the concept of a cultural journey – getting to know another culture through travel and in the process relativizing or expanding oneself through this new knowledge. Especially for small and medium-sized enterprises (SMEs), I recommend inviting an academic from a related field to come in and walk around the company. Since academics have a different way of thinking, he or she will be able to offer helpful perspectives.

Speaking of SMEs, how can these businesses in particular – which are largely responsible for the economic prosperity in Europe's rural regions – profit from university knowledge?

From my perspective, the key is for SMEs to get together and approach the university as a group. For many small businesses, going to a university to propose a research project is much too intimidating to do alone. As part of a fleet it's a different story.

Especially in regional development projects where lots of SMEs have settled, increased cooperation between academics and these local businesses can be fruitful, and they often work well too. This is something universities have recognized, and they have responded to this need. But it is still no easy topic because, for one, SMEs don't necessarily think in terms of cooperation. Here in Austria, it has been our experience that incentives are often necessary to get people to work together.

In addition to the already mentioned different points of departure often encountered in company–university cooperation projects, there is also sometimes a kind of resentment that stems from romanticized views. To put it polemically: on the one side you have the noble scholars and on the other a band of profit optimizers ruled by the forces of evil. Do stereotypes like these play a role among your colleagues?

Not as a stereotype and not to that extent, but sometimes there is justifiable skepticism. Business and academia embody different values. In all fairness, academics haven't always maintained ethical integrity in their research quests either, especially not if they saw some financial advantage. But one of the tasks of the university is to serve as society's conscience, it is expected to look at people more closely and critically than other systems might.

And there are also cases in academic practice in which a scientist might be working on a long-term research project and wants to be put under outside pressure to release yet unverified results. Or in the applied sciences, academics often hardly find the time to conduct theoretical basic research anymore because they are constantly trying to land third-party funding. We often forget that our sciences didn't start out as applied sciences. The early natural scientists studied the stars, there was nothing applied about that. And look where that has led!

You address a classic area of conflict in university–company cooperation: a company asks for a quick solution to a specific product problem and in response it gets a thick, detailed, painstakingly compiled research report. In group dynamics this would be called a "disturbance."

Exactly, instead of a quick solution to the problem sometimes good academics will give you a perfectly specified study more than 20 pages long, which can make some managers nervous.

But back to the advantages of two different mentalities joining forces: Churchill once said "protect the rebel," and Ruth Cohn's slogan "disturbances take precedence" became a fundamental principle in working with groups. These make sense if we constructively challenge agreement, especially from a technological, organizational, but also ethical standpoint. From this perspective, the disturbing person becomes valuable because he or she is the only one who introduces a different perspective, which is something everyone usu-ally senses anyway. A very uncomfortable process, of course, because we prefer to keep group situations nice and cozy.

When you begin a new project with businesses, what are some of the things you pay particular attention to?

We start by jointly defining the task in detail, and in the beginning this takes place verbally. Not until after the matter has been discussed and clarified do we draw up a written draft. Otherwise, before we even get started we have different assumptions that weren't arrived at through mutual agreement. Then, the time factor is important. If you want to work with academics, you need a different time frame and special organizational forms to go along with it. This doesn't just happen parallel to everyday business. I pay attention to time, organizational structures, and spaces for intensive exchange. And, particularly in the beginning, I am careful to make sure everyone is aware of the advantages and disadvantages of the collaboration. Never run from the disadvantages!

Between the ivory tower and the pressure to publish

In English-speaking countries, cooperation between universities and companies is a consciously cultivated practice. There, expertise on this subject has developed gradually and become quite sophisticated. In continental Europe this subject is relatively new. Sometimes it almost seems as if European universities in their ivory towers have only reluctantly begun letting down their golden hair, and only because of federal budget cuts and the need to raise third-party funding.

Because of immigration from Europe, North American society managed to successfully combine religion with economy. In the Calvinist tradition, money has different connotations than in the prudish Catholic mind-set of Central Europe, where capital was called Mammon and considered with disdain. This Calvinist influence can also be found in England and Scandinavia and tends to regard someone who earns well as one of the chosen!

In the United States sponsors don't think twice about funding research projects on obscure subjects and are even proud of this. America definitely has a two-pronged approach.

Federal funding cuts have had an enormous impact on university work in recent years. Today, if I advised a young academic – a historian or an archeolo-gist, for example – not to worry about anything, told him to do research on what moved him, and that everything would prove useful in the end, he'd look at me like I was crazy. Instead, he or she would be better off studying something else, publishing as much as possible, and giving talks at lots of conventions. These talks plus the publications are tallied and measured against other knowledge indexes. Today we have an academic system that plays by a lot of the same rules that govern competitive free market economies. It's a system based on a science-oriented paradigm of thought that values quantity over quality.

But aside from the financial issue, wouldn't this pressure to mingle with industry tend to improve academic activity?

Absolutely. In general, the permeability of the borders between systems is something I consider positive, and I can confirm this from personal experience. There is a tendency to cut oneself off from other systems, and today we must examine the situation retrospectively and ask ourselves if we haven't perhaps sealed off the borders of academic activity too tightly in the past, whether it is enough to let the privileged contemplate other people without giving the latter the same opportunities.

The ivory tower image is right if the impression we have of academia is that it is just answering its own questions, but that's not enough. In this sense it is a positive thing that economic pressure points to certain problems, but it doesn't solve them. Inevitably the question is whether or not a society that is as infused with commercial and technological values as ours – and this is unprecedented in history – is even capable of recognizing challenges that go beyond these considerations.

When cultural scientists collaborate in company research

You mentioned the borders between systems and your own experiences as a practicing philosopher in the real existing economy. According to your notion of transdisciplinary research and development, what do you see as being the new tasks of cultural scientists within the framework of, let's say, technology projects?

Let me give you an example. Under the theme "What is a Car?" VW commissioned us to do product research, and we invited various academics to come together for this purpose. **Everyone thinks he knows what a car is, but nothing could be further from the truth!** *A car is a major product that combines so many different needs and contradictions.*

With Volvo, for example, market shares hit rock bottom once because the safety engineers won an internal company battle against the designers and engine developers. They developed a safe family vehicle that flopped. A car is a contradiction between safety and danger. That's the way it has always been, travel is the balance between danger and curiosity. If the safety issue completely takes over, you lose large groups of potential customers. All of our major commodities are contradictory products. We have to understand this, and to

do so we need psychologists, philosophers, historians ... Take television, for instance. If you want to understand this product, go talk to a historian of religion and let him tell you about the omnipresence of God.

The problem is that at the present there aren't that many people in the position to ask these questions at all. And precisely these types of analyses plus perhaps the facilitation of technological-commercial development processes could turn into powerful new fields of activity for cultural scientists of all different disciplines.

That is exactly right! It would be kind of like grassroots anthropology because at some point in the history of mankind we made the collective decision to try to respond to our needs and contradictions with products. Earlier, that wasn't possible, so these needs were met by religion, for example, or myths and fairy tales. The world of commodities is essentially also a fairy tale or metaphysical world. Instead of turning the magic ring to be transported somewhere else ...

... we go online and buy a plane ticket with Ryanair?

Exactly! And as soon as we realize that products are an answer to human nature, they are no longer mere technological objects! That's a completely ridiculous one-dimensional perspective. Once you understand that all material things made by man are products that started out as ideas, it becomes easier to transcend barriers. But don't get me wrong, if you were going to ask if I think it's enough to satisfy needs with products, my answer is of course not!

What advice would you give an R&D-oriented company that was planning on incorporating the expertise of cultural scientists?

On the one side, the aim is to gain a better understanding of one's own products or services: Which demands does my solution really satisfy? In what environment will my product be applied? Going beyond our company mentality, what might be the reasons why this service didn't work? Without the answers to these kinds of questions, the knowledge of cultural scientists is useless. Dealing with these questions can help to break the ice on both sides. But most people don't believe that this kind of dialogue will really accomplish anything.

Ethics brings up a completely different aspect, and this has nothing to do with morality. By ethics I mean the examination of interdisciplinary issues: Is this okay? Is this the way we want it? Does this cover everything? Can we justify this? A vast field perhaps, but with the help of external facilitators, in this case philosophers, all that is needed is to establish places where people can come and ponder these questions together. The questions won't arise on their own because company process doesn't really provide for this. They might come up occasionally, but they don't get communicated, which is a big disadvantage.

On the other side, the universities – refer also to the beginning of this inter- view – should do more to come out of the laboratory and lecture halls and make their research results public. They spend a lot of time thinking about other peo- ple, but they never share their conclusions, or if they do, it is expressed as criticism through the mass media or in academic publications. That doesn't do any good.

It always comes down to the joint search for knowledge...
... yes, with the emphasis on arriving at this knowledge together. Disciplined science is, as the name says, disciplined, it rigidly adheres to its models and constraints, and ignores the fact that impurities are constantly invading the system. And companies are obsessed with the efficiency of their technological, organizational, or economic ways of thinking. The task now is to dig deeper and ask each other more questions, complement each other's expertise, and in this way continue to evolve.

Thank you for the interview.

Univ. Prof. Dr. Peter Heintel
Philosophie und Gruppendynamik
Alpen-Adria-Universität Klagenfurt
IFF – Fakultät für Interdisziplinäre Forschung
und Fortbildung (Klagenfurt–Graz–Wien)
Abteilung Weiterbildung
und systemische Interventionsforschung
Sterneckstraße 15
A-9020 Klagenfurt/Austria
Tel. (+43-463) 2700-6121
peter.heintel@uni-klu.ac.at

Cooperation Between Competitors:
Intercompany and Interuniversity Research Collaboration

Interview by *Renata Schmidtkunz*

Belgium

"The contract also requires companies to send their researchers to us. And we make them a part of the team, which of course also includes IMEC staff. But it's not as if the companies can sit back and wait until someone puts the research results on their desks. No, they are involved in the project through their resident scientists, which of course provides the opportunity for excellent communication."

Joint research – separate marketing. A carefully developed cooperation model between academia and industry in one of Europe's oldest university towns.

IMEC (Interuniversity MicroElectronics Center) with headquarters in Leuven, Belgium, is Europe's leading independent research center in the fields of nanotechnology and microelectronics. IMEC was founded in 1984 as an initiative of the Flemish regional government. The aim was and is to bridge the gap between basic university research and the needs of the micro- and nanoelectronics industry in Europe. That means that universities can continue to concentrate on basic research, while IMEC conducts applied research oriented on the technological needs of industry. In addition, by running this research center, it also hopes to strengthen local and regional industry and to network Flemish universities.

After 22 years, during which IMEC has been instrumental in setting up one-of-a-kind cooperation models throughout Europe between researchers and the microelectronics industry, the Center, a non-profit organization, is doing well: 68% of the annual budget of roughly 200 million euros comes from global industries that conduct and sponsor research at IMEC; 22% is financed by domestic corporate sponsors; 8% of the budget is financed by the EU, and 2% by the European Space Agency.

IMEC employs roughly 1,400 people from different European and non-European countries; 500 of them are so-called "industrial residents," i.e. researchers who have been commissioned by industrial partners to conduct research at IMEC, as well as guest researchers who have chosen IMEC's excellent equipment and infrastructure over those of other research centers.

The IMEC mission is to keep ahead of industrial needs by three to ten years in R&D conducted in the fields of microelectronics, nanotechnology, design methods, and technologies for ICT systems (information and communication technology). Constant contact with its industrial partners ensures that research stays focused on industry needs.

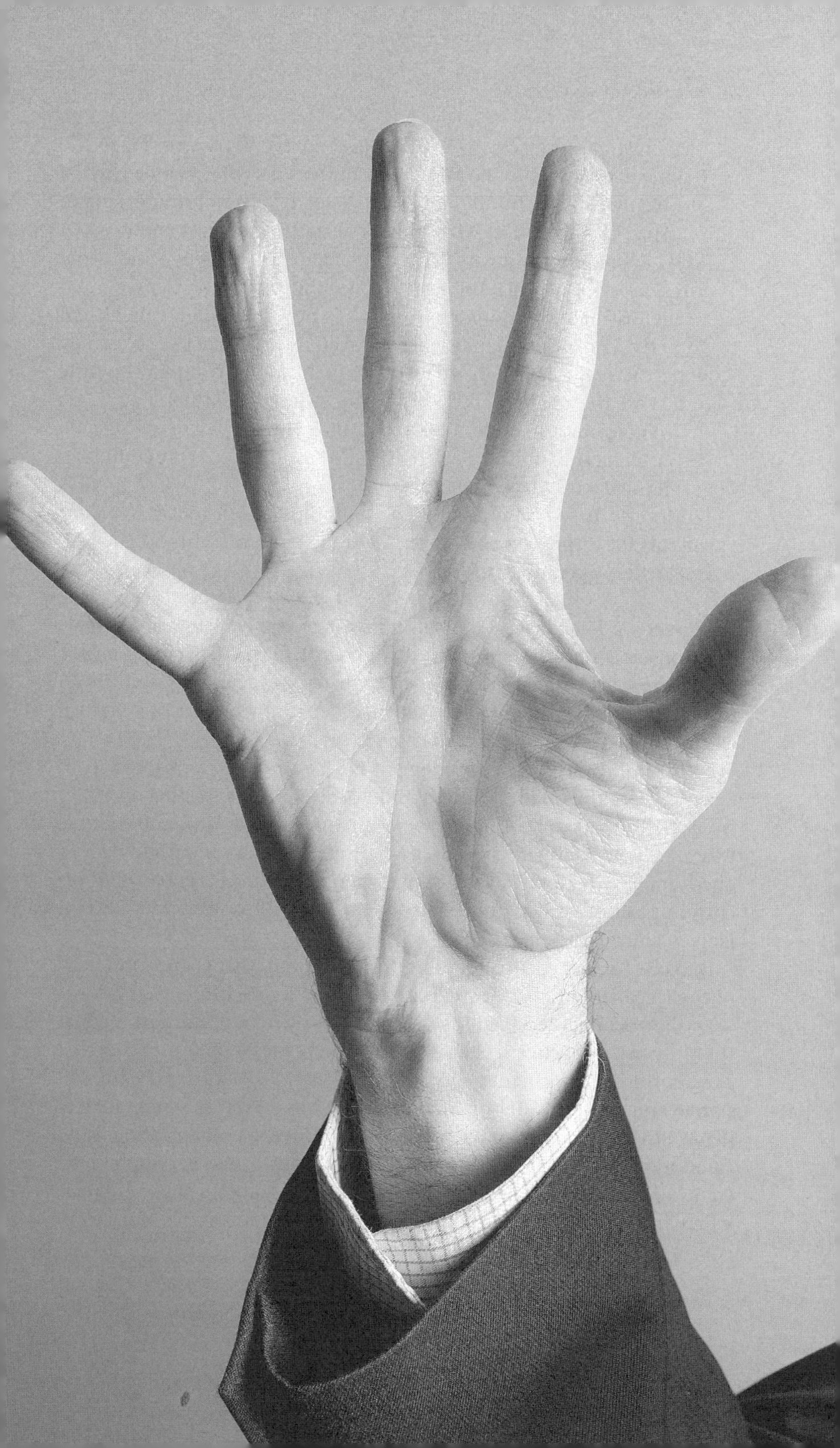

The research projects, which last three to five years on average, are designed according to global market surveys on what will be needed and wanted in the future and cover fields of research important to industry. Collaboration with industrial partners comes about when IMEC approaches industry with a research plan or when industrial enterprises with special interests or needs approach the Center.

One distinctive feature about IMEC is how it handles intellectual property. The results of jointly undertaken research can be accessed by all partners, as clearly defined by contract; however, company-specific results are the exclusive property of the industrial partner.

This cooperation model makes it possible for companies in the same sector to conduct joint research even if they are fierce competitors.

IMEC currently has nine industrial "core partners." One of them is Eindhoven-based NXP, a subsidiary of Philips that produces semiconductors for IC process and device technology. At present, NXP has 60 researchers operating at IMEC.

Dr. Roger De Keersmaecker, vice president of the Division of Strategic Relations, is a member of the Executive Board of IMEC. He is responsible for all top-level interaction between industry, academic research facilities, and the government. De Keersmaecker is a professor of electrical engineering at the University of Leuven.

Dr. John Schmitz, vice president of NXP and director of research in Leuven, started working for the Philips Research Labs in 1984 and later went on to work for a number of companies in the USA and Europe, where he was chiefly involved in designing processors and highly developed semiconductors, always in managerial positions. He has held his current position as director of research at the Philips subsidiary NXP in Leuven since December 2005.

Dr. Lode Lauwers is director of strategic program partnerships and head of the Department for Silicon Process and Device Technology at IMEC.

Dr. Jan Wauters received his doctorate in nuclear physics in 1993 at the University of Leuven. He has been working for IMEC since 1996, first as scientific editor, then as coordinator of IMEC's strategic training initiative and strategic advisor. Since 2005 he has served as senior advisor of the Roger Van Overstraeten Society, a nonprofit organization dedicated to conveying a basic understanding of electronics to people in all walks of life. In 2006 he founded the Flanders NanoBusiness Alliance, where he currently serves as director. At IMEC he is in charge of industry communication and heads the marketing division.

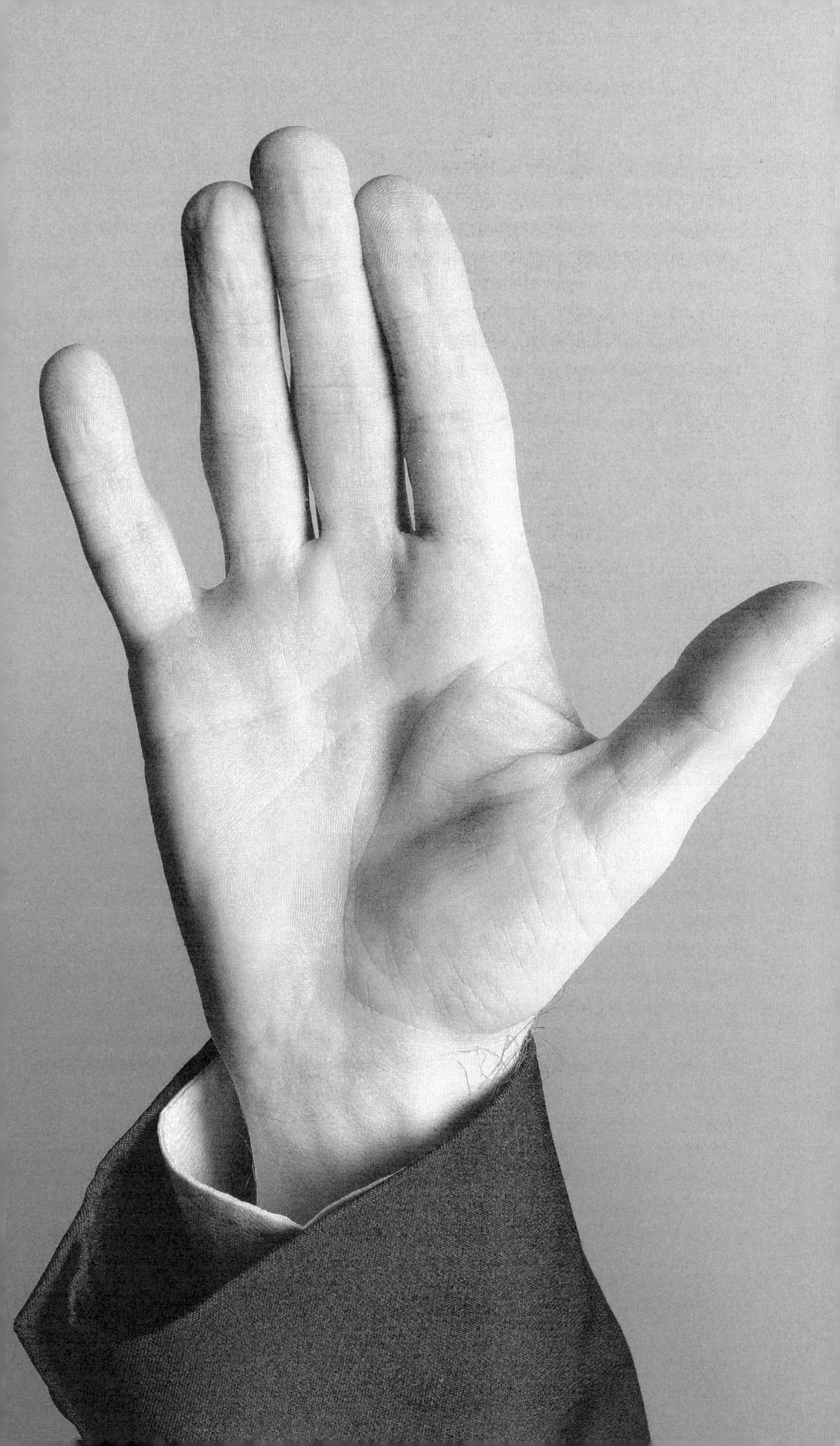

Dr. De Keersmaecker, what are the biggest concerns, what do you consider the heart of the collaboration between IMEC and its industrial partners?

*ROGER DE KEERSMAECKER/IMEC: Our industrial affiliation programs are large-scale collaborations. At the moment nine companies are involved, all of them global players in the computer and communication sector. Our goal is to anticipate demand in the next few years and to develop highly sophisticated process modules in the field of micro-silicon technology. Participating companies share their know-how and are free to take home any results from their joint research they might need to develop their own technologies. The research teams consist of IMEC researchers as well as what we call resident scientists who are sent to us by the companies. **Our cooperation model is based on the companies' willingness to share the expenses, results, resources, and of course also the risks of these large-scale research programs.** When you invest in a research process, there is no way of knowing ahead of time whether the specific advance one is working on will turn out successful. If you have to make a decision on limited resources, you might make the wrong decision, whereas if you conduct your research with others, you have more options and that also means reducing your risks.*

IMEC was founded in 1984. Is it true that at that time public authorities were no longer providing adequate support for research? What was the political background for starting a research institute like IMEC?

Public knowledge is usually generated at the universities. This is where research results are published and a major part of this is financed by public authorities, that is to say by the government. IMEC is a facility positioned somewhere between the universities, where basic research is conducted, and industry, which needs applied research. The idea behind IMEC was to create a place where knowledge that is generated at universities can be developed into useful information for application in industry. We are a kind of intermediary between general university knowledge and the specific knowledge required by industry. Our task as we see it is to translate problems that we have encountered through our contact with industry into more fundamental questions that might also fall under basic university research, and later to put together all the results in such a way as to render them useful to industry.

JOHN SCHMITZ/NXP: What De Keersmaecker says is absolutely true. Maybe I should add that industry has a global technology road map describing the common problems and questions in certain industry sectors. So, all the players in a certain industry sector are confronted with the same problems. And now,

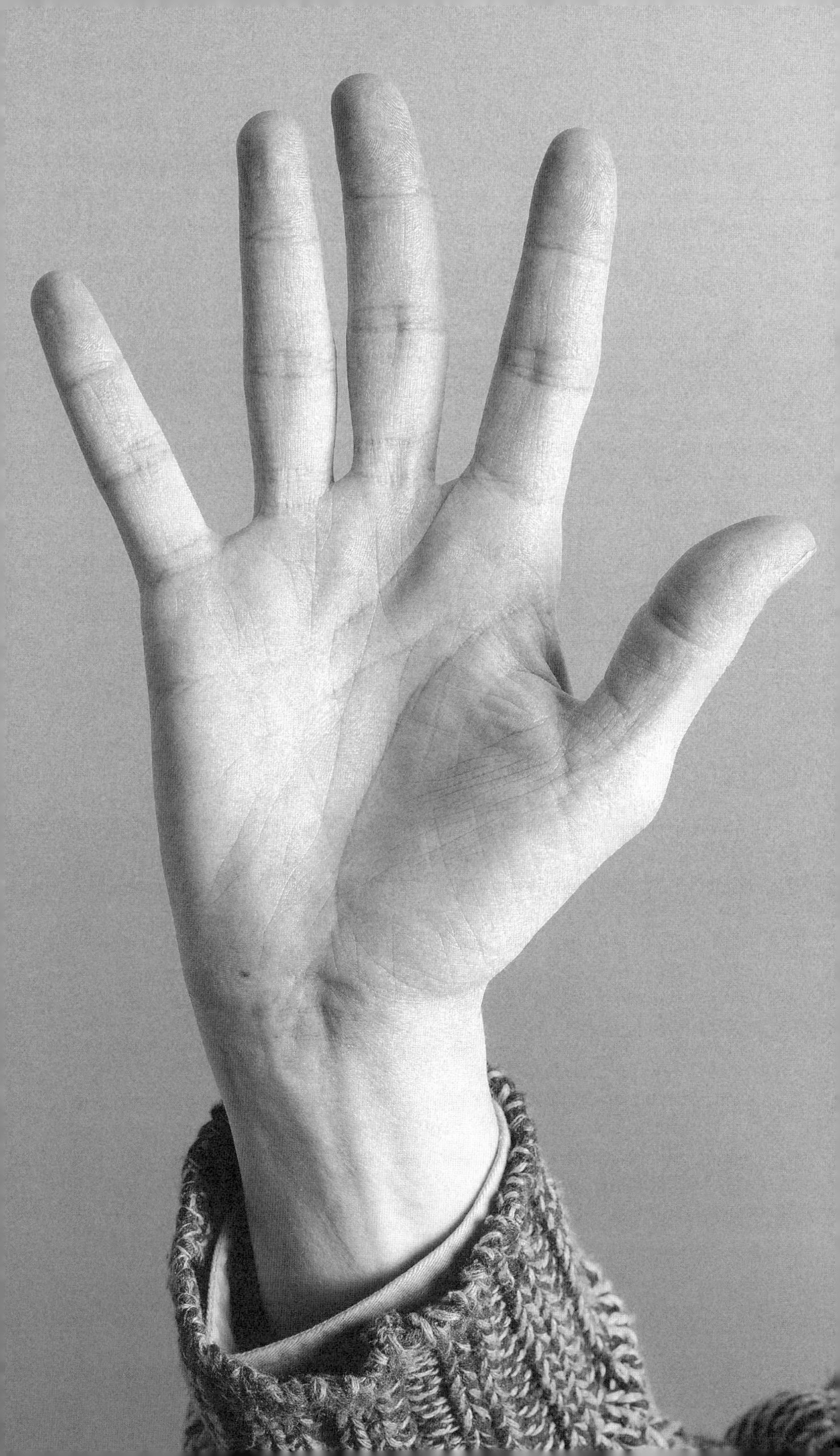

*you as an individual company can either try to solve the problems alone or you say: Let's solve these problems together! Like De Keersmaecker said, large industrial corporations are faced with enormous problems and challenges. **Studies show that the gap between basic research and the application of research results in industry is increasing.** That is a negative trend. Industry has therefore come to the conclusion that the only way to get a handle on this problem is to conduct applied research together. And that is exactly what happens at IMEC.*

At the universities, of course, there is a lot of basic knowledge being generated, but that doesn't mean that this basic knowledge can also be used directly in industry. And with IMEC we, the nine companies participating in the research projects, have the opportunity as a group to ask what the big problems and questions are, and then to draw up a research plan with the engineers from IMEC and our own experts.

Nine companies split the costs for an enormous research program. That lowers not only expenses but also increases knowledge. You can imagine that all the researchers involved will bring their knowledge into the project and make it available to the group. We are sharing not only the costs but also know-how. And that makes this form of cooperation a very strong concept indeed.

But aren't the partners participating in the research projects market competitors?

Yes, absolutely! Samsung and Etel and my own company are fierce adversaries competing in the same segment of the electronics market. But in order for our products to be successful on the market, we have to develop certain tools. Tools we can use together. I think that is the best word for it, the golden word. Let me give you an example. For producing electronic chips we use the same equipment or the same tools, and despite that we still all make different chips. How each of us uses the tools that we develop together is a company secret, we don't share this with each other. And that is the key to the success of our collaboration.

So it doesn't mean that rivals can't work together?

No, on the contrary. And our collaboration demonstrates this again and again. On the market we are extremely fierce rivals, but we can still work together in many regards. Together we can do things that would be extremely costly, or even too expensive, for us to do alone.

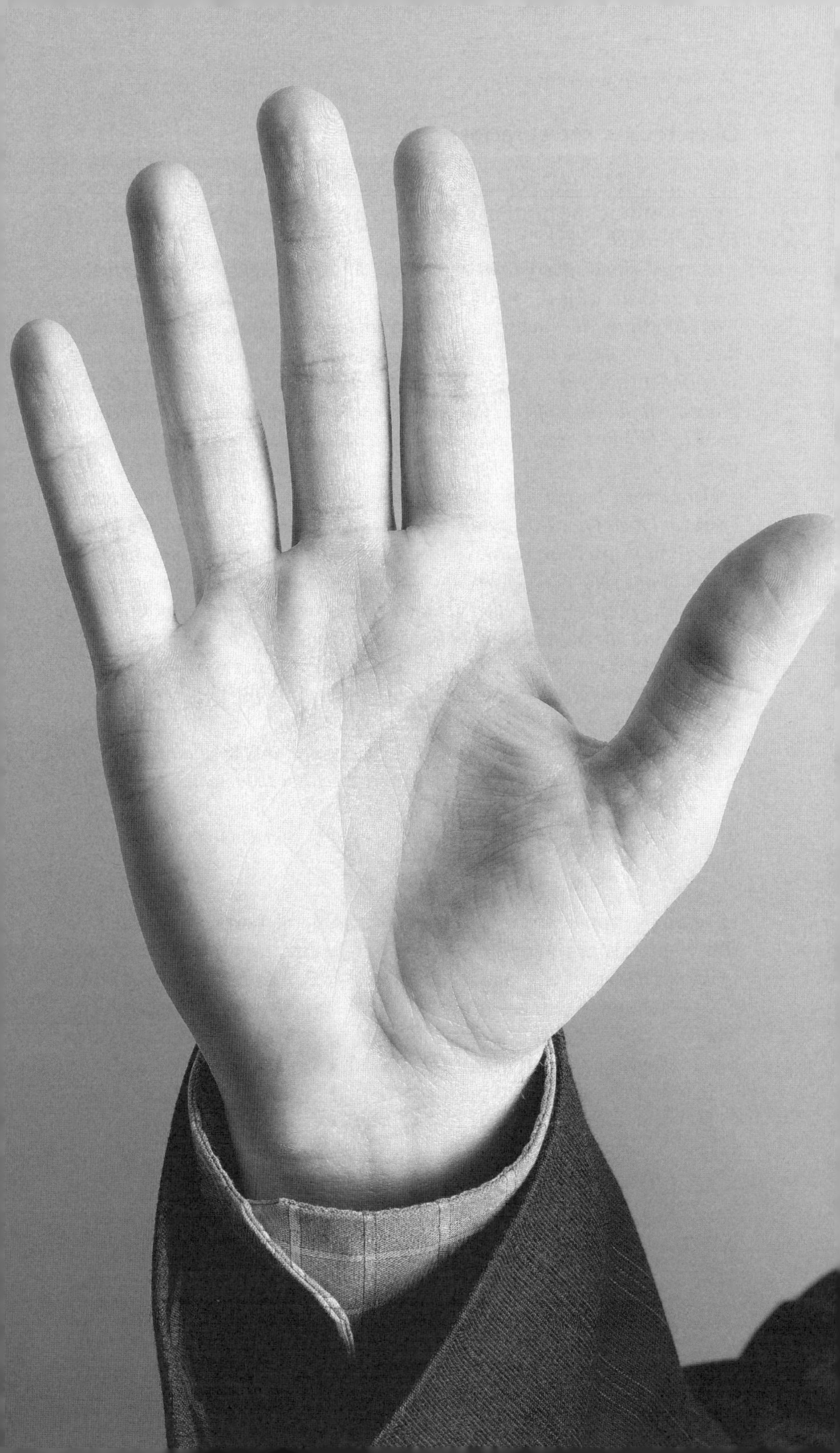

Open results, secret recipes

How can IMEC convince its industrial partners that those research results that constitute the heart of a company's production won't be passed on to the competition?

It's very simple. Every company decides for itself what it wants and doesn't want to share with the other research partners. Each company knows best what constitutes its intellectual property and what it doesn't want to share. In practice there are never problems in this area.

LODE LAUWERS/IMEC: *We don't expect the companies to share their intellectual property (IP) in the project. That is a very important element of our cooperation model. The joint research program is intended to operate as a self-sufficient unit.*

Here or there the participating companies will have to contribute their experiences and their knowledge, either informally or formally. Because just keeping your IP to yourself isn't always productive. And sometimes you profit more if you share your experiences. But the cooperation model is set up so that the research program develops its own IP on the basis of the joint research results. And the participating industrial partners have the right to do whatever they want with those research results. Often we don't even know why one of our partners incorporates the research results into his technology because we don't know what his actual IP is. What we offer are only small pieces of information, each partner sticks to his priorities, and by maintaining constant contact with our partners we try to ensure that the research project stays focused on their needs. Because if one day one of our partners decided he wasn't interested in the project anymore, we'd have a serious problem on our hands.

Listening to you describe the essence of your collaboration, I thought of this comparison; you tell me if it fits. I'm a restaurant owner who wants to cook a delicious chicken dish, and you – John Schmitz, in other words NXP – are also a restaurant owner on the same street of the same city, my rival, and you want to cook fish. Of course your aim is to do better business with your fish than I with my chicken.

So now, both of us go to the market – in this case that would be IMEC – to buy tomatoes, which I use for my secret recipe and you for your secret

---

Photo   (**see page 35**) Roger De Keersmaecker, Vice President of the Division of Strategic Relations at IMEC

fish dish. Thus both of us would profit from IMEC without needing to disclose the secret of his great recipe to the other person, the competition.

JOHN SCHMITZ/NXP: Yes, you could look at it that way, or maybe more like this: we both want to cook, and what happens at IMEC is that through joint research we develop the best ovens, in other words the best conditions for everyone, so that you can cook your chicken and I can cook my fish.

LODE LAUWERS/IMEC: Of course it is very important that we sign a contract in advance with every partner – and there can be up to 25 partners working on any given research project. For this reason, much has already been settled at the start of the project, including how much of his or her IP each partner wants to contribute and what he/she is prepared to share with the other partners and with IMEC.

Mr. Schmitz, what do you and the other partners consider special about this cooperation model? Do you have a basis for comparison with other research facilities and cooperation models?

JOHN SCHMITZ/NXP: In the kind of research we do, the main thing we need, if we want to be successful, is a high degree of intellectual potential. And we need knowledge from very diverse areas. There are 1,400 people working at IMEC and some 800 of them are academics. They possess excellent, first-class knowledge, which we, for our part, need in order to understand exactly what the materials required for production are composed of and how they react. And you won't find this high degree of knowledge in too many other places in the world. That's the most important thing.

On top of that, the IMEC research laboratory is excellently equipped and always kept at the cutting edge. In our line, even in research you have to have the best that technology has to offer, you can't operate with outdated equipment.

And then, each company profits from the excellence of the other industrial partners. These three elements, that is, IMEC's intellectual potential, the cutting-edge scientific technology, and the advanced level of the partners, taken together this is what makes this cooperation model so unique. You won't find this in too many other places either.

Photo   (**see page** **39**) John Schmitz, Vice President of NXP and Director of Research in Leuven

Cooperation agreements: aims, contracts, expectations, and their limitations

Does that mean that from your perspective as an industrial partner there are no difficulties? Is everything pure bliss? Don't you have to make compromises sometimes, accept things that aren't necessarily what you want?

Of course there are things that you have to learn to accept. When you're married, you are no longer alone and you have a responsibility toward others. You have to get used to your partner's idiosyncrasies. And that's why it's very important to carefully define the content of the research project and to be clearly aware of what you are willing to accept about the other person. **What you want is to come to a consensus with your new partners – all at the same time.** *Of course we have our road map, which means there is strong agreement about our common goals. But the details are the subject of constant debate. I would say that the content of our research program coincides roughly 80 percent with one's own expectations, and 20 percent to a lesser extent. Then when the results are on the table, one has to share the newly created IP with one's partners. Of course, that's still better than having no IP because then you can't do anything at all. As a company we came to the conclusion that this form of cooperation doesn't keep us from conducting our business successfully; on the contrary, it's a very good deal because the magic word in the whole equation is cuts: you cut your resources and your costs enormously.*

Do you have expectations that can't be met at all or maybe only partially?

As far as IMEC is concerned, our only expectations are that the Center provide us with the necessary infrastructure and intellectual know-how. And like in every good marriage, sometimes you have to rethink and discuss certain issues. But in the end you always arrive at a solution. No, I think that at the moment our expectations are being met.

How does IMEC see this from its point of view?

ROGER DE KEERSMAECKER/IMEC: I think John hit the nail on the head: you have certain expectations and sometimes particularly at the start of the project they don't really coincide. For example, maybe we want more money or more guarantees from the companies, more flexibility, and more freedom. But all

---

Photo   (**see page 43**) Lode Lauwers, Director of Strategic Program Partnerships and Head of the Department for Silicon Process and Device Technology at IMEC

these things get cast into a contract. Every once in a while you have to make readjustments because the world isn't perfect. Sometimes it is necessary to pare down your own expectations in the form of a contract. Often we wish our partners would invest more time in our research. That would be one of our expectations.

LODE LAUWERS/IMEC: I think that our reciprocal expectations are pretty much the same. We expect our partners to feel seriously committed to the programs. To me this almost goes without saying because the research programs are very expensive. We hope the people responsible for the budget within the participating companies are all making sure that the funds are being used as efficiently as possible. And our partners expect the same thing of us.

When we approach a company, we try to make absolutely clear which services we can and cannot provide. We stress, for example, that we aren't a company and have limited resources. And sometimes we have to make clear in the middle of a project that certain things simply aren't or are not yet feasible. Meanwhile we have found that our industrial partners – such as NXP, which has been our partner the longest – are very understanding. And by now, if questions do arise along the way, we also know that our partners won't come to us with impossible demands. Both sides – IMEC and the partners – know what can be achieved and what is unrealistic. So I see the matter of expectations as very positive.

The company: partner, client, guest, or . . .?

Do you sometimes find yourself caught in a moral dilemma? Say one of your industrial partners asks you to conduct research that goes against your ethics, for example war-related research?

ROGER DE KEERSMAECKER/IMEC: Some time ago we decided against any participation in military research projects. Our researchers demanded this clear stance from us out of fear of becoming involved themselves or perhaps already being involved without having been asked. And that is why we made this ethical commitment, but in effect it is no different than a kitchen knife: you can use it to peel tomatoes as well as for less honorable intentions.

And how is chip technology used? On the one hand you need them for all portable devices – telephones, computers, etc. – on the other, they are also used in tanks. We don't control how our research ends up being used. But out of consideration to our researchers and because of the ethical commitment we

---

Photo   (**see page 47**) Jan Wauters is in charge of industry communication and heads the marketing division at IMEC

made, if a potential partner came to us with a research request and it turned out he had been commissioned by the defense department, we would say: no thanks.

If you can't control or define by contract how something ends up being used, then it is more or less a matter of trust?

LODE LAUWERS/IMEC: Yes, a matter of trust, but we don't work with partners we don't know. And besides, our IP programs are all still many stages away from becoming any kind of final product.

In this cooperation constellation between the companies and the research operation IMEC – who would you say is the slave and who is the master?

ROGER DE KEERSMAECKER/IMEC: Well now, that's a very good question. (grins) We prefer – and I think John Schmitz will concur – to speak of partners rather than clients.

Earlier, John Schmitz used the analogy of marriage. Allow me to expand on this idea: at the beginning of a marriage the couple might think they are partners too, but as time goes by, they may realize that one is the slave and the other gives the orders.

LODE LAUWERS/IMEC: I don't think any marriage will last long without arrangements as to how the couple wants to live together. For example, if I'm a good cook and you are a good handywoman, then I'll do the cooking and you can take care of the practical things. But you have to have some kind of agreement – and that's what we do here too. At the onset of every research project it is very important to us to remain independent and make sure none of the industrial partners dictate what we do.

Afterwards, we meet with our partners and through attentive listening we try to find out exactly what they need. We have very good contacts with industry and the research and environmental scene, and in this way we gather information and determine how technology will develop, what will be needed in the future. And then we discuss that with our potential or already existing partners.

Then we draw up a research plan and present it to the companies, who then sign a contract with us. The contract also requires companies to send their researchers to us. And we make them a part of the team, which of course also includes IMEC staff. But it's not as if the companies can sit back and wait until someone puts the research results on their desks. No, they are involved in the project through their resident scientists, which of course provides the opportunity for excellent communication. Of course IMEC has to produce a usable result in the end. And at times a partner can become pretty nervous and might point out that he paid a lot of money, so we should work faster. Like in every good marriage there are also tensions, but they never get out of hand.

That means that at the end of the day you are offering industry a service?

Yes. One of our core competencies here at IMEC is to set up and execute programs of this scale. Ten years ago we had far fewer partners. In a lot of ways it was much simpler. Now we hold a conference twice a year where we present the progress of the research project to our partners. And then one of them might say: Gentlemen, I think you have to readjust your course to the left or right. Then I go back home, check to see which of the partners' suggestions are feasible, and let them know. During a given program we have to correct our course at least twice a year to make sure the partners stay satisfied and don't abandon ship.

Company culture and open research

Does that mean that the cooperation models have to be readjusted from time to time too? Haven't you had to correct it a few times since 1984, perhaps because of lessons learned from the joint projects?

The concept of a shared IP has been part of the model since the eighties – and I think we were the first to design this type of program. The step of integrating the partners in the strategic developments and decisions didn't come until five to seven years ago. And 2000 was a decisive year because NXP became one of our main partners. Since then we have been involving the leading IC companies in our strategies and with their help predicting what industry will be needing in the next three to ten years.

That means you have to know your partners better than they know you?

No, the companies have to know us well too. Because we don't operate in the market, they do. Other mechanisms are at work there – and these need to be discussed again and again.

I assume there is a difference between the internal processes and structures of the companies and a research institute. How do you manage to harmonize these differences?

JOHN SCHMITZ/NXP: Yes, of course we have different structures and processes. But in the end we come together as human beings. We have our researchers here, and sitting on the other side of the table are IMEC's researchers and project manager. And this is where the interaction takes place. The people working here come from different parts of the world and sometimes there are cultural differences to be dealt with, but that isn't really a problem because in the end what we are basically discussing are technological problems – and the language of technology is a universal one.

LODE LAUWERS/IMEC: As a research institute IMEC doesn't manufacture any products, whereas with companies, production is precisely what it's all about in the end. Our natural communication partners at the companies are therefore research directors, like John Schmitz. But over the years we have learned that there are much bigger "mechanisms" at work than the research departments, namely the development department and production itself. Most of us here at IMEC started out as researchers. That's why we had to learn how to decipher the messages our partners were sending us. And we did this by listening closely to what they said. In the end what it always comes down to are the results arrived at together. And despite limited resources and time we all want to achieve the best result possible for our partners. And if you fall short of that, you won't be punished. You just have to reorganize . . .

ROGER DE KEERSMAECKER/IMEC: Over the years we have definitely undergone a change in thinking. The fact that we chose the cooperation model described above has forced us to take on an industrially minded mentality because this is one of the key expectations of our industrial partners – and we discuss this daily. They expect us to be a super professional organization and not a university one. By growing and involving our partners more and more over the years, we have come to accept this.

And what has also brought about this change in thinking is the fact that the proportion of funds coming from the government has decreased and that our budget is financed largely by industry.

To what extent have the demands of industry – e.g. to come up with a usable result by a certain deadline because a new technology has to go into production – had a fundamental impact on the nature of your scientific work?

In the beginning, of course, there was this typical academic Trojan Horse wariness. **But since then, we have come to realize that collaborating with industry doesn't necessarily mean that we can't conduct open research anymore.** *Our research just has to be more focused. Of course, we have also changed our way of researching . . .*

What mistakes occur most often in this kind of collaboration?

JOHN SCHMITZ/NXP: I would say that one of the main difficulties is under-estimating the size of a problem. For example, in the course of a project we might realize we're going to have to invest more resources, or that we allotted too much time for a project and should have focused on fewer points instead.

ROGER DE KEERSMAECKER/IMEC: I would say another problem is the right com-munication. We used to think it was enough to have an industrial partner representative here, and she or he would transfer the information to the right people at the company. We discovered that this isn't always the case, and we have had to establish communication structures at both the labor and manage-ment levels. Today, we go about this in a much more structured way. First, the engineers have a week of intensive brainstorming and discussion. Then their conclusions are used to brief the managers, who meet to discuss this internal information the following week.

Before we can proceed, we have to make sure all the partners understand the conclusions so that they can decide how to continue the project because this is a decision that requires everyone's approval.

Do you hire professional communication experts who "speak" the lan-guage of industry and academia and can act as facilitators, or do you rely on the communication skills of your colleagues and co-workers?

We haven't hired a facilitator yet. Over the past 15 years we have learned a great deal, and I think that we have got better.

When we started out, we had two or three industrial partners. Communi-cation was easier back then. But with nine globally operating companies who all have very pronounced egos you have to come up with other communication structures. As mentioned, that was one of the things we underestimated in the beginning, but now we are prepared.

JOHN SCHMITZ/NXP: This brings us to an important point in this collabora-tion: as an industrial partner you have to make sure that the internal paths of communication are structured in such a way that the research results are not only communicated properly but also implemented. Every one of us invests quite a bit of money into these projects and if the results are not implemented properly in the companies, it is just a waste of money. This is another process that can take years – and something we underestimated in the beginning.

We talked about the researchers and resident scientists here at IMEC – people from different countries and with very different cultural back-grounds. How good does your understanding of human nature have to be in order to successfully conduct this type of cooperation project?

ROGER DE KEERSMAECKER/IMEC: As with every other management that tries to achieve a task through people and their work, we too have to know what makes people tick in order to know how to get them to do what we need them to do. We have people here from 50 different nations. And that means that we have to be aware of the cultural differences, today perhaps more than ever before. Twenty years ago we mostly had Belgian scientists working here, and once in a while we might have got a Dutch or French researcher, but today we have some 100 scientists from Holland at IMEC, maybe 30 from the USA, 50 or 60 from France, 40 from China, 20 from Japan, and so on. We had to develop an awareness for the cultural differences, and a few years ago we even started increasing consciousness for this among our employees.

Do you follow a given strategy, or do you just act on instinct? After all, how much time does one have to deal with these delicate matters considering all the scientific and technical issues that have to be resolved as well?

If you don't deal with these matters, the project might not work properly, so you are forced to think about it and do something. We organize evening events where we invite external consultants to come and tell us a little about how other cultures operate, for example in Japan: how people there think, communicate, deal with confrontations . . . And we also organize theme evenings with, for example, Korean music and dance.

Do the industrial partners take an interest in these events?

JOHN SCHMITZ/NXP: Let me put it this way: the better people feel, the better they work. That's why it's very important to us to implement these kinds of programs.

ROGER DE KEERSMAECKER/IMEC: I'd be lying if I said we didn't have any problems, that we were a perfect organization. Of course problems of a cultural or personal nature are going to arise among the researchers, but it's important to keep in mind that we're dealing with scientists, that everyone here shares a passion for science. They want to make discoveries and develop new technologies, and be proud of what they have achieved.

But if someone really has a personal problem, he or she can consult a trusted IMEC staff member. With the employee's consent this third-party intermediary will try to solve the problem. On the company side there are team leaders who are responsible for this. We only offer this service to our own people; it's up to the companies to provide counseling for their personnel.

Over the past 20 years what has changed overall in the way you manage human resources?

Ten years ago we still expected a scientist to have an "international profile." We expected scientists from Belgium to operate just like the ones from Germany, Spain, the USA, or Asia. But over the years we realized this wasn't so. There are very different approaches to communication, hierarchy, and authority: that's something we also have to learn how to deal with. For example, if we had a conference on Friday and made a decision, we assumed the matter was settled. But what sometimes happened was that a Japanese or Korean colleague would come in on Monday and say: I thought about this over the weekend, and I think we should do it differently. I used to regard this kind of thing as a sign of undependability. Now I know that it has to do with different communication patterns. And that's something we have to adapt to.

Global research and regional development

Dr. De Keersmaecker, to what extent does the region profit from the collaboration between industrial partners and IMEC?

For one, we continuously incorporate local companies into our research projects. Moreover, we are not just interested in international partners. That right there is a plus for the region. In addition, we can sell results from research projects to local and regional companies too, thanks to our IP policy. As mentioned earlier, results from the large-scale research projects belong to each of the cooperation partners – and at the end of the project each individual can do with it as he or she pleases. The companies, for example, develop microprocessors or mobile telephones. And we at IMEC are free to pass the same know-how on to local companies, which in turn use it to develop something else. But only if there are no contractually stipulated restrictions, e.g. that the joint research can only be used elsewhere after a period of one year. But that's more the exception to the rule.

But how do local companies know what you are currently working on? Do you approach companies because you know they could use a given technology, or do the companies come to you?

JAN WAUTERS/IMEC: We communicate with local companies. The companies themselves profit most because 60 percent of our engineers and scientists take jobs with local industry after they finish their research projects with IMEC. That's the best kind of transfer. We have joint research programs with some of these companies, but if the companies are too small or don't have research departments, we assist in the knowledge-transfer process and help implement the research results and new technologies in their companies, and we also assist in the production process. That applies especially to small and medium-sized enterprises because often there is no other way for them to conduct research.

In this respect we are kind of a go-between. We observe the technological trends and developments and pass our observations on to the regional companies, so that they can take part and develop new products.

IMEC is a nonprofit company. Does this mean you can pass your know-how and research results on to local companies at a low price because you are not forced to make money?

ROGER DE KEERSMAECKER/IMEC: Yes, that is true, but one also has to understand what "nonprofit" means. It's not that we aren't allowed to make a profit, but we don't have shareholders behind us demanding that we do so. We have to reinvest the money we earn, which means it comes back to IMEC. We use it to finance new equipment and for new research programs. But you're right, turning a profit is not our main concern. We try to invest 5 and sell for 50 and in the end we should break even. So, if something costs 100, we should shoot for earning 110 because you've always got unexpected expenses. And if the project ends up costing 102, then we end up making a profit of 8, which we can invest in other projects.

How do you determine the price of research results that you sell to local and regional companies?

JAN WAUTERS/IMEC: Here in Flanders we have the IWT, Institute for the Promotion of Innovation by Science and Technology. Together with regional companies it defines what constitutes an innovative project, and the IWT makes co-financing proposals for projects – on the local or EU level. In other words, the IWT is a partner in research projects and the companies receive co-financing from government agencies.

We at IMEC, however, are not just concerned with research and technology transfer. We try to provide for every aspect of a successful technology industry. We have a training center and we support entrepreneurship. In addition, we have a capital fund to promote start-up companies. We also start subsidiaries, spin-off companies that will go on to become independent enterprises in the future.

In conclusion, what is your advice to people interested in participating in the kind of cooperation project practiced at IMEC?

JOHN SCHMITZ/NXP: To me the two most important things are, first: you have to have a common problem and the will to solve it jointly. And everyone involved in the cooperation project has to understand and recognize that the other partners' problems are also their own problems.

And second: once the problem has been solved, it's up to each partner to make sure that the results are implemented in his or her own company.

ROGER DE KEERSMAECKER/IMEC: My first bit of advice is work hard!

But what is also important in this context is having a region that will give you long-term support – like here in Flanders. In our 22 years of operation, the proportion of our budget covered by state funding has gone down, but we still continue to receive support – "from a distance," so to speak. The Flemish government has never stuck its nose into our business. It sends its people to make sure we deliver what we promised. But it never interferes and always lets us implement our own strategies. I consider that very important.

*Second, it's important to operate globally. We are a global industry. From the start, we worked with companies in the USA and Japan and built up our credibility. We have always worked with the best in the world. **Being good in your region is fine, but ultimately you need a certain segment where you are the best in the world.***

And you also have to be selective. Part of our success is our focus. Narrowing down your goal to a few things you want to be good at. Once you've achieved that, you can expand. But I think it's a mistake for companies to want to accomplish too much all at once. Of course, if you've got 1,000 employees, you're going to be able to do more than if you just have 200. But you have to go about it in an intelligent manner. And if you are very good in a few things and not just pretty good in everything, then potential partners will come looking for you. Of course, the network is very important too. With 500, 600, or 1,000 partners spread out all over the world that's a significant source of information. And last but not least, what's most important is making sure you yourself are at the center of that network.

Thank you for the interview.

Key project insight

- Research collaboration with companies and/or competitors within the same sector is an extremely interesting service offered by a neutral "university" platform. It requires an intensive process of working out the rules of the cooperation project as well as the rights of use of the resulting intellectual property.
- Long-term financial and nonmaterial support from the state.
- Networking with regional SMEs: Global research, cooperation with the leading companies in the field, plus a conscious process of advising and passing on the research results to local small and medium-sized enterprises.
- Structured, continuous communication processes between IMEC research staff, the future users at the companies, and management!
- Company researchers work with their university counterparts at a "third site": resident model.

- Active community building and the promotion of intercultural cooperation. Sensitizing the researchers to different culturally determined styles of communication.
- Think big! You only achieve credibility in international cooperation by doing business with the best partners. That requires accurate knowledge of the global research scene and the industrial market.
- Flat hierarchies ensure speedy reaction and simplified communication.
- A good working climate leads to greater success for the company; therefore being observant and mindful of employee satisfaction is important for increasing employee effort.

Scientist, Inventor, Entrepreneur?:
Cooperation and Technology Transfer I

Interview by *Michael Prellberg*

Germany

EMBL
Main Laboratory
Main Entrance
Mech. Workshop
Kinderhaus

"I am a researcher who loves the balance between the academic world and concrete applications. I think it is extremely important for institutions committed to basic research to demonstrate that what we are doing here has a concrete application for concrete people."

From discovery in the laboratory to the launching of a successful product: patenting, licensing, cooperation with industry, or founding your own company. EMBLEM helps researchers get from the ivory tower to the market – and back again.

Heidelberg is one of the hottest places in the world; scientists here call it a hot spot for basic research in molecular biology, and the Molecular Biology Laboratory (EMBL), which is financed by 18 EU countries, draws them like a magnet. Since molecular biology has begun setting the course in biotechnology, concerns all over the world have begun paying close attention to what goes on in the Heidelberg laboratories.

In 1974 EMBL was founded as a European research facility for basic research in molecular biology. Eighty percent of the approximately 750 employees come from 18 member states; the other fifth, from the rest of the world. They conduct research in some 80 different units, of which two-thirds are based in Heidelberg. EMBL has branch laboratories in Grenoble, Hamburg, Hinxton, and Monterotondo.

EMBL Enterprise Management, abbreviated EMBLEM, was founded in 1999 as a 100 percent EMBL subsidiary with the goal of marketing intellectual property. As opposed to many similar companies, EMBLEM is not concerned with merely selling licenses to industry. Its aim is to combine inventions, scientific consulting, and collaboration into useful packages and in this way offer its customers one-stop concepts.

The future belongs to the field of biotechnology. Nobody will argue with that. It's just the present that still refuses to submit.

The enormous expectations have yet to be met on every level. So far, biotechnology has remained a promise. That is one of EMBLEM's major problems: it must live with the fact that the initial euphoria has come and gone. Pharmaceutical companies no longer shell out money for licenses and patents by the handful.

In this all but stable situation Martin Raditsch, head of business development at EMBLEM, supports EMBL researchers Joe Lewis and George Reid in founding their own company: Elara Pharmaceuticals.

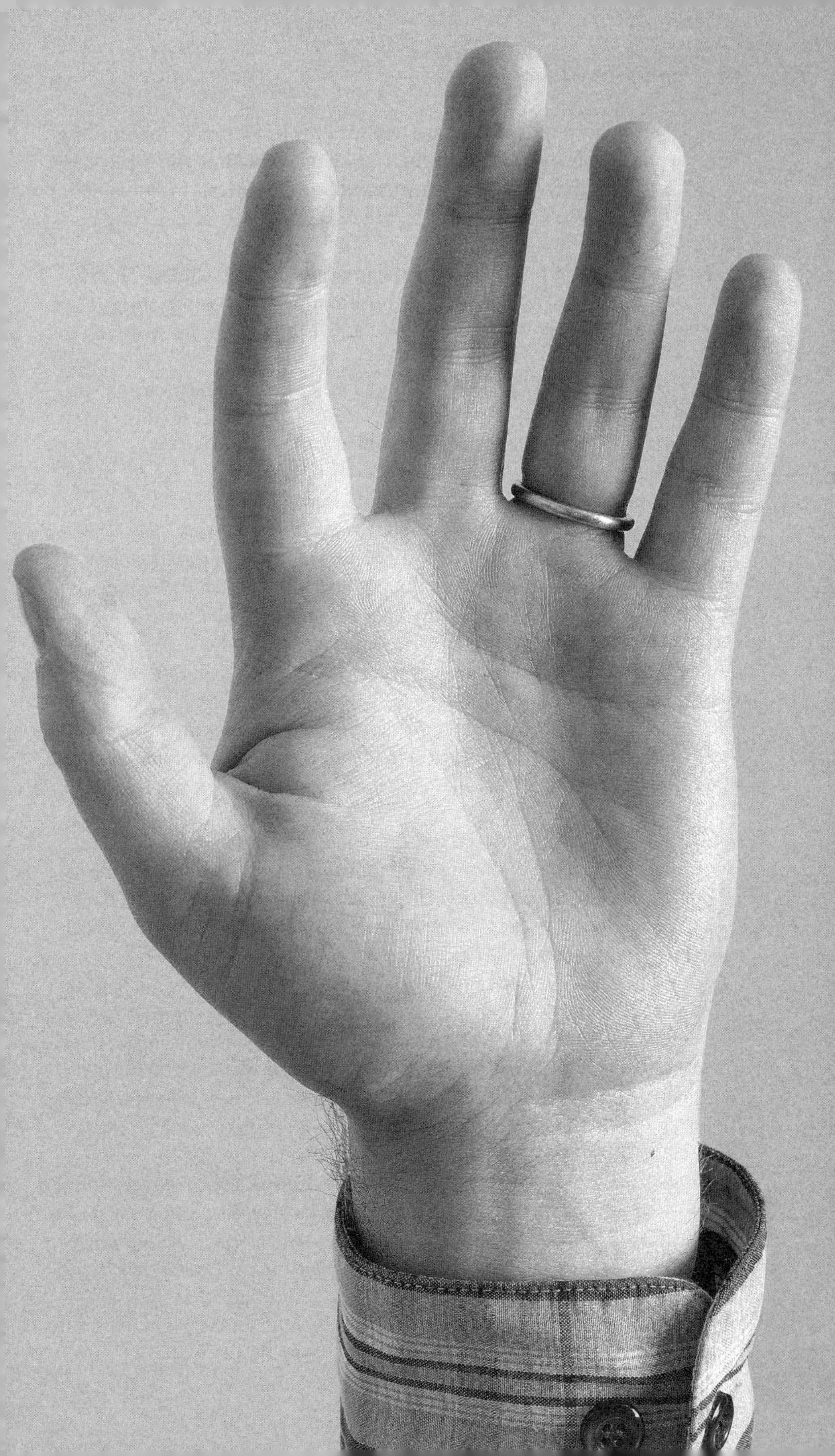

Dr. Joe Lewis, born in Scotland, earned his PhD in Vienna before finding his way to EMBL for the first time. He then worked at the Wellcome Trust Centre in Edinburgh and for Anadys Pharmaceuticals in Heidelberg. Lewis went back to EMBL in 2004, where he has been researching ever since.

Dr. Martin Raditsch is deputy managing director at EMBLEM. He studied biochemistry in Heidelberg and chose a molecular biology subject for his dissertation. After working for BASF for several years, he switched to EMBLEM in 2001.

Dr. George Reid was born in Scotland. He worked for the pharmaceutical group Pfizer, among others, before switching to EMBL. Reid has been managing director at Elara Pharmaceuticals since August 2006.

Do good molecular biologists make good entrepreneurs?

JOE LEWIS/EMBL: I suppose what you're actually asking is aren't these researchers really idealistic nut cases? No, they're not. The three founders of Elara – George Reid, Frank Gannon, and myself – are not only familiar with the world of laboratories and lecture halls but have also had years of experience in and with industry. Our CEO, George Reid, for example, used to work for Pfizer. That means we have insight in both the business and the academic worlds.

GEORGE REID/ELARA: The immense advantage that our experience in the business world gives us is in our being able to judge what potential our research results might have in practice. And we want to exploit this potential.

Is that why you have chosen a two-pronged approach?

It isn't as if we set out to develop something marketable. We started out with a purely academic research approach: we were looking for an answer. But of course we are shrewd and experienced enough to realize when we stumble onto something that can be applied in practice.

JOE LEWIS/EMBL: Researching at EMBL is always about basic research. That might seem worlds away from concrete applications. But appearances can be deceiving – basic research is what innovation is built on. And what we have discovered is innovative. If we achieve our goals, there is already demand for our findings in practice. What we are working on could help wipe out breast cancer. And our chances of success are frankly quite high.

Why has EMBL taken the responsibility of helping Elara in this way?

MARTIN RADITSCH/EMBLEM: It's a fairly new development. Until the mid-nineties EMBL's notion of research was so strict that scientists had to justify their work to secure financing for patent submissions. Patents stood for

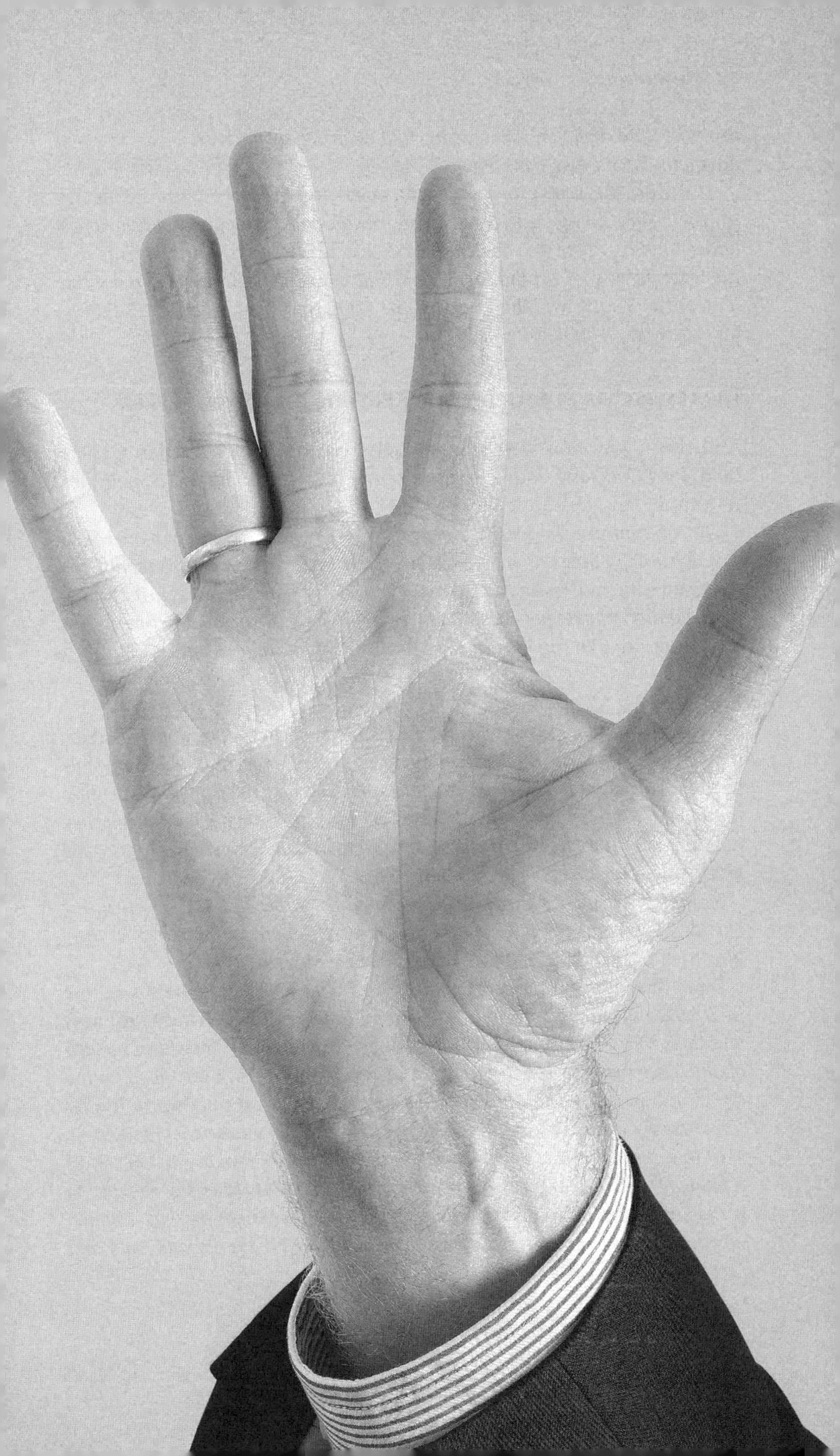

usability and therefore supposedly had nothing to do with basic research anymore. This image has changed.

At EMBL the people in charge sat down together to discuss technology transfer. They acknowledged two fundamental facts: they didn't have enough understanding about the subject matter, i.e. about licenses and patents and the legal paperwork behind them, and that it wasn't their job to acquire this knowledge. That's why they founded the 100 percent subsidiary EMBLEM as an expert for technology transfer.

The researcher as entrepreneur? Passing the entrance exam

And that's why you were immediately sold on the idea when George Reid, Joe Lewis, and Frank Gannon came to you, excited about their idea to found Elara?

On the contrary. In general we tend to be skeptical. Whenever an EMBL researcher approaches us with the idea of starting his or her own company, we do our own internal research based on a formal assessment process to determine patentability, market potential, and probability of success. **And before we do anything, we ask the researchers to draw up a business plan.**

Why that?

First, we want to find out whether the scientists are in the position to describe the purpose and aims of their company in a way understandable to people outside of academia – whether the applicants can make the leap from scientific thinking to thinking in business terms. For many, this is a big hurdle, but they have to get past it because if they can't communicate their aims in this early phase, they won't be able to do it later either.

Second, a business plan can check euphoria: is the idea still so attractive outside the laboratory? For the first time researchers are forced to look at their idea from the outside. That can have a healthy, sobering effect.

Some researchers wander off optimistically but never come back with the final business plan. It turns out to be too much work. It makes them deal with questions they've never had anything to do with and which they have no idea about. They find this annoying. And that kind of thing isn't very good for the ego either. Someone who throws in the towel and says: It's not worth it after all! probably wouldn't have much staying power as an entrepreneur either.

In that sense drawing up a business plan serves as a filter: if you don't want it enough or don't have a really convincing project, you either say goodbye to your plan altogether or at least reconsider seriously. And that's the way it should be because only someone who has thought about all the ifs and buts and

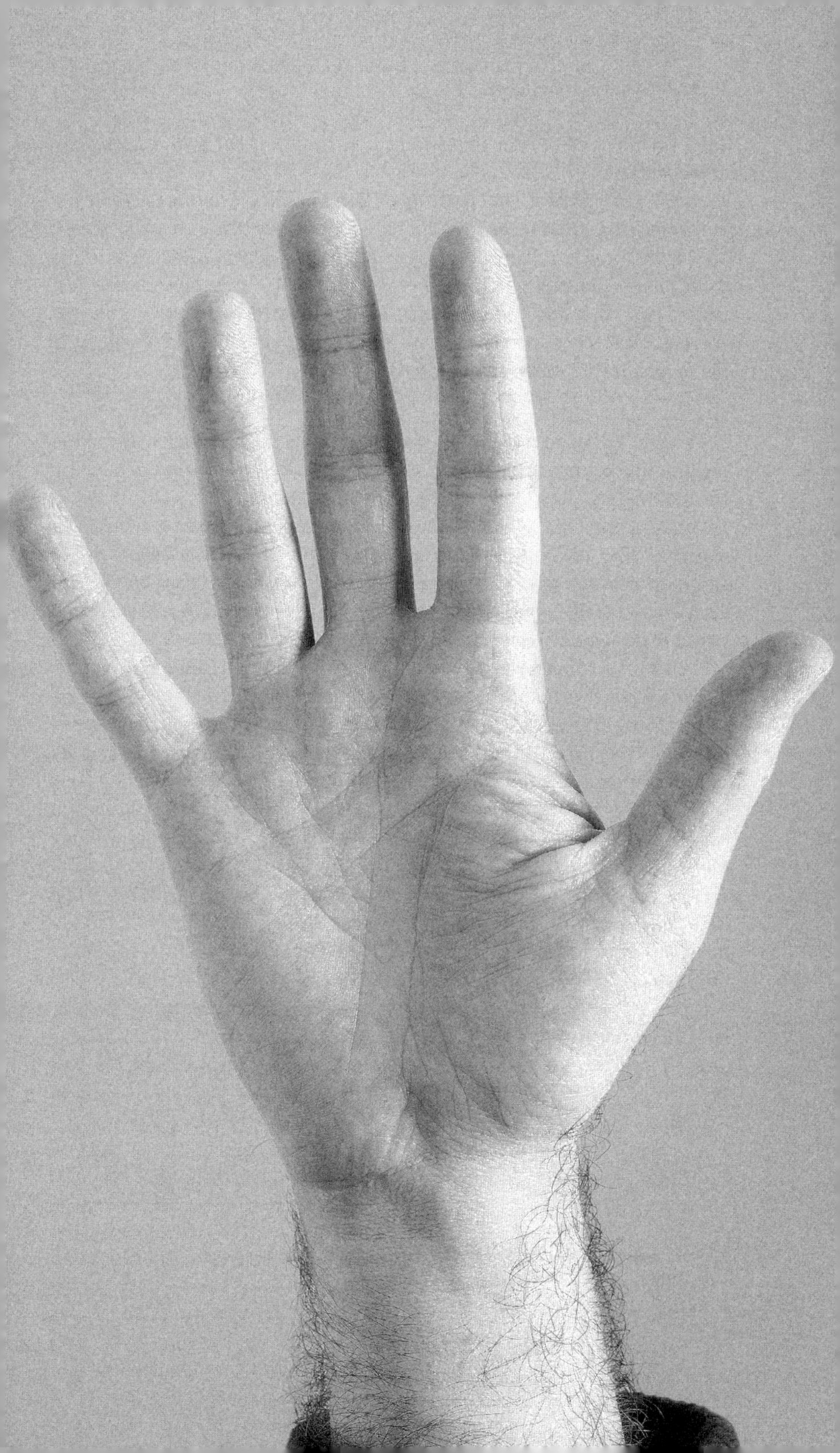

still wants to go through with it will be able to stick it out with his company on the market.

GEORGE REID/ELARA: I can't agree more: these business plans are definitely a pain in the neck. Getting the scientific part down on paper was easy enough, we knew what we were talking about there. But then they wanted our assessment of the project's market potential, they asked us how much money we needed and for how long, and had we thought about where the money was supposed to come from? They wanted to know our strategic options. And we sat there and looked at each other: strategic options?

Does EMBLEM even have the expertise to judge both the scientific value and the market potential?

JOE LEWIS/EMBL: Definitely. That's all they do all day …

GEORGE REID/ELARA: … and they still don't rely exclusively on their own expertise. They also consult independent experts. There are a few experts in the world who can assess our work – and they do not keep their opinions to themselves. I really appreciate this outside expertise. It gives me feedback and helps me navigate my course: where do I want to go?

There's a lot of money at stake here. If we were hoping EMBLEM was going to give us money just because of our pretty blue eyes – forget it!

MARTIN RADITSCH/EMBLEM: We don't make things easy for ourselves. In the beginning you asked if molecular biologists made good entrepreneurs. My answer is: some of them. In this case we were lucky that we were dealing with scientists who had enough professional and life experience to be able to put themselves in the entrepreneur's shoes. It's always a question of wanting something enough.

Once you know you want it, the next question is, is it worth it? A start-up isn't founded as an end in itself, it's supposed to create value. If corporations are interested in their know-how, they should have to pay a high price for it.

Before we seriously think about launching a new company, we look at how much rise in value we can expect. We've done this a few times already and I'd say we are more experienced at it now. So we decided the best thing to do was to start a company, invest in it, and give it two years to pay off.

Photo   (**see page 71**) Joe Lewis, born in Scotland, has been researching for EMBL since 2004

Start-up strategies

Why don't you just keep conducting research for those two years and not bother with starting a new company?

JOE LEWIS/EMBL: Because it's much faster this way. It's a time issue. We'd lose too much time otherwise. Here everything is already set up and we can just continue our research. And what we can't handle ourselves, we outsource. These things all cost money.

Whose money?

At the moment we are getting seed financing from EmblVentures, EMBL's own venture capital investor. We discussed the subject at great length until we finally accepted that in this case it was worth founding a separate company and that it would be supported financially by our internal fund. That is enough to get us to the next phase of our research. Once we have reached that level, we'll be interesting to other investors specializing in venture capital. And with their funding we will be able to get through all the preclinical tests.

Besides EmblVentures, isn't anyone else interested at the moment?

GEORGE REID/ELARA: It's a tedious task to find someone to give you money.

JOE LEWIS/EMBL: We started thinking early on about our different license, cooperation, and start-up options, and after thinking long and hard we decided to take it into our own hands, at least until we got through the preclinical phase.

Once we got to that point, we could collaborate with a pharmaceutical or biotech corporation, and they are the only ones who can afford to finance testing in the expensive clinical phase.

*MARTIN RADITSCH/EMBLEM: Corporations wouldn't be interested yet, it's still too vague for them. They wouldn't be able to assess how sustainable or imple-mentable our approach is. To the pharmaceutical industry our whole project still has project status. We've given this a lot of thought, precisely these three options. **But it wasn't far enough along for a license or a partnership. That just left the third option: to start our own company.***

JOE LEWIS/EMBL: We shopped around, but it's extremely hard to find an investor for the phase we're in. At the moment seed financing fits for Elara – and still involves plenty of risk. Fortunately, we have a first-class advisory board. People like former head of research at BASF Pharma Erich Schlick not only bring with them an excellent network, but their names can also help impress

---

Photo   (**see page 75**) Martin Raditsch, Deputy Managing Director at EMBLEM

investors. People like that don't join an advisory board unless they sense a strong potential.

How does one recruit that kind of heavyweight support for one's advisory board?

Through the network.

MARTIN RADITSCH/EMBLEM: *If you're being backed by an institution like EMBL, which has 30 years of experience in research and a reputation to match, the network almost builds itself. EMBLEM and EmblVentures latch seamlessly on to this. I'm not trying to be pretentious, but EMBL is a hot spot, one of these places that really encourages and supports research. And of course people are watching to see what a George Reid or a Joe Lewis are working on. It may not be a particularly big network, but it's very tight-knit.*

Does that mean the biotech companies know what research work is being conducted at EMBL?

At least we don't have to run around advertising ourselves. I'm afraid that wouldn't be enough anyway. When scientists get together at conventions, they talk shop, and that's how researcher A hears from researcher B what researcher C is working on. These informal channels work very well and often have access to amazingly current information: 80 percent of all contacts are made through this network of scientists. From out of the blue we might get a call from someone we've never talked to who will proceed to tell us about the exciting work our EMBL researchers are doing.

In other words, if someone comes to EMBLEM, he knows exactly why?

In two out of three cases people have very concrete, very specific expectations. They know exactly what they are talking about and what they want. We certainly don't send out unsolicited e-mails to companies hoping someone will take the bait. That's not necessary because we already have our own network with industry contacts. Everyone knows each other. That doesn't leave much room for coincidence.

Do other research facilities and universities network in the same professional way?

It takes several years before it reaches that stage. A lot of institutions thought they could generate an income stream relying mainly on technology transfer, but that's much too short-term a way of thinking.

Photo   (**see page 79**) George Reid is CEO at Elara Pharmaceuticals

At EMBLEM we were able to pay off our setup and operating costs in five years. Many German universities have similar goals, but without the proper funding and support of high-level management they are destined to fail. At many universities technology transfer is a one-man show run basically on good intentions. And then people expect money to flow freely with immediate results. A paradoxical situation. It can't possibly work out.

The "nonstop shop" for technology transfer

How does one do things right? In other words, if you were put in charge of technology transfer at a typical university today, and you could implement whatever you wanted, what would you do?

Assuming the university is serious about its commitment to technology transfer, I would favor a system like EMBLEM. A closed technology transfer unit that included the areas of scouting, patents, administration, and business. In addition, a technology development fund enabling us to push projects in the direction of commercialization. Employees should all have experience in their specialty fields: this makes getting started much easier. To facilitate start-up projects we'd need an on-site incubator or technology park or we'd have to build these, and finally, we would want to launch an associated seed fund.

That's a pretty long and compact wish list. Before we dissect it, why would you want an incubator? I thought the idea of the incubator as an innovation accelerator had turned out to be a flop.

To me, an incubator, in the most pragmatic sense, simply gives a new spin-off the first on-campus roof over its head.

Let's continue along these pragmatic lines. You mentioned scouting: so now, what's a scout?

A scout is like a truffle pig. His job is to systematically go from one project group to the next, actively attend seminars, listen, but also talk to the scientists – introduce himself. People need to know who he is and what he does – know him well enough to want to approach him too. A scout is like an ambassador to an internal system.

You can put a postdoc in charge of this, but it's generally better to get someone with more experience and a fuller picture of the market. At a larger university one scout isn't enough, you need a team of researchers. But if you're dealing with a manageable group of 400 scientists, one scout should be sufficient.

You also mentioned a technology development fund. How do I get one of those?

The first thing I need is launching aid – for this, my target contacts are federal or state governments. Currently, there are programs available that financially bridge the gap between academic projects and industrial research.

Couldn't the seed fund you mentioned earlier, in other words private investors, perform the same tasks as this technology development fund?

It could, but that's not its responsibility, and unfortunately it is very difficult to get private investors excited about university projects. Seed funds usually don't start to play a role until later at the spin-off stage.

But without the financial launching aid you mentioned nothing happens.

An initial investment will obviously be necessary because as far as I know the perpetuum mobile of technology transfer hasn't been invented yet. Then as technology transfer starts to return a profit, these revenues can be reinvested in internal technology development funds. The goal should be for the technology transfer to feed the fund. In no case should research funds be used here.

And yet the university is supposed to foot the technology transfer bill?

No, that's not the intention. The investment must be paid back in the long term. In the beginning, of course, you want to avoid promising financing for longer than ten years. That also serves as a kind of stimulus: if the grant will run out in three or five years, you have to present your progress in a convincing way: We put the money to good use. Look how far we've come already. We'll be able to stand on our own feet in the foreseeable future.

Isn't it easier to just have the university allot you a sufficient annual budget?

Yes, absolutely, provided the university was willing. Very comfortable, but it also makes you lazy: we've already got our funding, there's no reason to expend more effort than necessary. In that sense independence is more demanding. That's why I'm convinced that this model is more likely to produce the desired results.

And how long does it take?

It doesn't take long for the investment to yield a return, but you shouldn't expect to make a profit for the first eight to ten years; after all, the goal is to make a long-term profit. The fact that EMBLEM was able to support itself after just five years of operation is a rare exception.

Do you think that the reason EMBLEM works so well is because its structures aren't yet as rigid and ossified as those of most universities?

I don't think it has to do with the structures themselves. What does guarantee success is when the university or the institute fully backs the idea of technology transfer, rather than just using it to please politicians and aid organizations. If this basic requirement is met, technology transfer can be successfully implemented in every structure.

And the scientists play along in every structure?

There are research facilities that own all the marketing rights and can license them whenever and to whomever they please. It's different with EMBL. No one is forced or in any way required to market their results. That would tremendously undermine and destroy the spirit of open research and collegial exchange.

How do researchers even learn about a technology transfer opportunity and the services EMBLEM offers?

Through regular conferences and seminars we show EMBL researchers what technology transfer means, what it is worth, and how it is conducted. At the same time researchers get a feel for what can be marketed, what patent criteria is, or what makes a promising start-up. Through these efforts we can get people to put their reservations aside.

Does that mean all researchers are open to your work?

Not all of them. Roughly every fifth researcher considers commercializing his research results as selling out – in those cases we always get a strict: no.

That's something I can't completely understand because the purported separation between "pure" science and "commercial" industry doesn't really exist. Even if it's something basic researchers don't like to hear, half of all postdocs end up in industry. Those are the statistics, and there's nothing wrong with that. For us there is an enormous advantage to this because, as a result, our contacts don't just spread in the academic world but also to the high echelons of the pharmaceutical and biotech corporations. We know them and they know us. This is our network.

JOE LEWIS/EMBL: And it's exactly these contacts in the pharmaceutical industry that we are going to use for Elara by tapping the network we built up through EMBL.

Researcher coaching – business consulting

Just a second: even though Elara doesn't even want to use them yet?

We don't see things that categorically. You always have the option of entering into an appropriate form of partnership.

Why does this only work with corporations? What's to keep someone from establishing a partnership with a smaller company?

Nothing. The company would, however, through its past record have to demonstrate and ensure that it was in the position to finance expensive research over an extended period of time.

But, if I understand correctly, you're not going to find a company like that anywhere in the world.

I wouldn't go so far as to say that. There might be some somewhere, and then there would be no reason not to form a partnership with them. But to me the decisive factor is whether they have the strength and the resources to fund our research until the end. I can't imagine anything worse than to have to give up right before we reach our goal just because the partner runs out of money.

But isn't that the fundamental problem in basic research, that by nature the road to application is relatively long?

True. And that's why we need the venture capital investors. If you ask other researchers, they all say: Steer clear of venture capital! But they all need money. And where is it supposed to come from?

MARTIN RADITSCH/EMBLEM: To get your hands on that kind of money you have to start companies, but actually, with us, starting new companies isn't the rule. For that you need the right technology, the right team, enough money – and the conviction that this new company is viable. To date, we have only given a handful of companies this form of launching aid. We only take action when we see that the researchers want it and the investors want it. We don't talk anyone into it. If we think the concept is viable, that's when we get to work.

Only ten companies have been founded with EMBLEM's help over the past few years – that's not a lot compared to what others have to show. But I'll tell you something, not one of our companies has gone bankrupt, and that's something I'm proud of.

Not yet.

Sure, it's bound to happen, but this sector has had to get through a few tough years, and none of the companies folded: that's no small achievement.

How did you manage that?

As I mentioned before, we tend to be skeptical, almost reluctant at first. We don't automatically think it'll work out. Instead we examine all the cons, and unless there is a "but still," we remain wary about getting involved. This prudence pays off later.

GEORGE REID/ELARA: When we came to EMBLEM, we knew we had something special. But, of course, that didn't mean EMBLEM was immediately going to be as enthusiastic as we were. They thought about it for a long time: Where is the market? Who is going to be interested in this? Who can use it? Who is doing similar research? At first this skepticism really put a damper on our enthusiasm, but when we got the thumbs up, it made us feel all the more satisfied.

And then with EmblVentures we went through the same thing again. They completely took our results and plans apart, refused to take anything for granted. That was a hard reality check – but it was helpful. It made it irrevocably clear to us that if we wanted to be entrepreneurs, we had to be able to survive on the market. This wasn't a game.

What happens when the project doesn't stand up to the reality check? Does it get tossed?

MARTIN RADITSCH/EMBLEM: Then there really are just two options. Either the project is immediately discontinued – if you're lucky you can get it licensed first – or outside experts are called in and they take over the project. The latter, to be honest, is more of a theoretical possibility. That has to do with the fact that the scientists who have brought the project this far are the ones most familiar with it. It won't work without them.

That doesn't mean that outside experts can't strengthen the team after six months to a year. Once the company is running, the founders are more likely to give up entrepreneurial responsibility.

Are you really willing to do that?

JOE LEWIS/EMBL: I am a researcher who loves the balance between the academic world and concrete applications. I think it is extremely important for institutions committed to basic research to demonstrate that what we are doing here has a concrete application for concrete people.

What form of assistance did you expect from EMBLEM? If scientists start a company, they're probably not too eager to deal with all the red tape involved. Suddenly the tax authorities want information, then there are legal issues and new forms to be filled out every day. Isn't that when most scientists say: I'm done playing entrepreneur?

Unfortunately. None of us feels like weeding through fiscal law or learning the legal tricks of the trade. But the nice thing about EMBLEM is that we aren't left to fend for ourselves. EMBLEM has sent a number of companies into orbit before us and knows how to keep them from crashing back down to earth.

We have profited enormously from this. We don't have to bother with finding out what the right legal form is, inquiring about capitalization, or where and how and as what to register. Of course we could have done all the research ourselves. But if you've got someone you can trust, and he tells you this and that is best, but if you want we can also do this and that – that takes an awful lot of pressure off you.

It also eases your fear. It's new to us to think of ourselves as entrepreneurs. All of a sudden we have to deal with so many things that have absolutely nothing to do with our research, absolutely nothing to do with our goals for Elara. That can get you sidetracked.

And I'm gradually starting to realize that this is perfectly all right. EMBLEM shoulders a large part of this work for us, but it doesn't withhold anything from us. It's not as if EMBLEM was running the company, and we could calmly go on doing our research. No, they make sure we face our responsibilities: You want to be entrepreneurs? Then behave that way. Here we have this problem – how do you want to deal with it? Here we have another problem – what's your opinion? Decisions aren't delegated, they remain our responsibility – as entrepreneurs. EMBLEM takes over when it comes to implementing the decisions as smoothly as possible. It's the perfect division of labor.

GEORGE REID/ELARA: What we need at the moment is assistance with all these legal and fiscal matters. It's just not one of our fields of expertise.

JOE LEWIS/EMBL: And we aren't interested in turning it into one either.

But that's part of being an entrepreneur.

Yes, true, that's the problem. We accept the fact that we have to roll up our sleeves and deal with these matters, but for now it's nice to have a little guidance.

MARTIN RADITSCH/EMBLEM: Just a moment – I see a potential misunderstanding here. It's not that George or Joe has to understand all the tax issues. They can spread this responsibility among other people within the board of directors, but the board of directors is in charge of running Elara. EMBLEM doesn't have an active role in managing Elara. Period. That is a very important aspect. Our role is different, we give tips and advice when asked. And we help in any way possible. In other words, we're more like advisors.

That doesn't mean that our advice is always taken. Over the years there have been a number of instances when EMBLEM and the start-up company didn't share the same opinion. That's only natural. What is important to us is that the scientists are not only willing to tend to their own company but also demonstrate that will.

"Tend to" sounds so mild-mannered. I picture a scientist sitting in the passenger seat studying the car manual. At some point the founders are going to have to take the driver's seat themselves.

JOE LEWIS/EMBL: At EMBL we are in the privileged position of being able to completely dedicate up to 20 percent of our time to the needs of our company, whereby a workday will frequently extend into the evening.

That means for four days a week you are scientists researching for EMBL, and on the fifth day you're entrepreneurs doing research for Elara?

We're entrepreneurs every day. And we do our research for EMBL. But on the fifth day we attend to all the unpleasant business we mentioned earlier: tax matters, forms, applications, etc.

MARTIN RADITSCH/EMBLEM: At the moment we are still very important to Elara because the company is still at the start-up stage, and at this stage the demands on us in assisting the start-up are particularly high. In one or two years, when the investors are considering a second or third round of financing and the managers are discussing what course to steer the company onto – when the focus is on partnerships, fusions, sales, or even going public – we will have much less sway. Our influence decreases over the course of time. It's the natural process if EMBLEM's goal is technology transfer. And this transfer ultimately also means away from EMBL and away from us. For us as mother institute that simply means as soon as they are big enough to stand on their own two feet, we have to let them go.

JOE LEWIS/EMBL: The idea of becoming independent didn't seem all that attractive at first because we mostly considered the problems: Who is going to give us money? How much will it take us off-track from our actual research goals? Once we realized these problems could be solved, we became more optimistic. Now we have begun to dream at Elara, for example of a fusion with a pharmaceutical company.

Thank you for the interview.

Key project insight

- Universities are increasingly searching for new sources of income. The idea of marketing the results of university research, therefore, seems attractive – and lucrative. However, it can be very disappointing if marketing revenues don't flow but only trickle back to the universities. EMBLEM shows how technology transfer can be successful:
- Don't rush: it can take years to establish contacts with industry and to build these into a network. In the beginning, the most logical

approach is to tap the network of scientists. Many of them go on to the business world in the course of their careers and can open doors for you.

- If you're going to do it, do it right: a lot of times whoever is conducting technology transfer just seems to be going through the motions or sending up a test balloon. If there aren't enough financial and personal resources, a single individual will often end up trying to hawk the product on his own – with expectably poor response.
- Full support: it is impossible for technology transfer to yield good results as a one-man show. It must be outwardly and inwardly evident that marketing is important at all levels. Clear proof of this are regular events about technology transfer.
- Expertise please: if you want industry to take you seriously as a business partner, you need expertise. And researchers who are open to marketing their research results think the same way. Marketing without the necessary expertise is doomed to fail.
- Always be skeptical: if you try to turn everything into money, you'll end up running a junk store. On the other hand, if you are critical from the start and set high standards, you will ultimately only market that which has the potential for success.
- What goes around, comes around: if your own quality control is strict enough to filter out the half-baked ideas from the start, the result is high-quality service, which in turn raises the company's reputation in the eyes of its industrial business partners. Eventually, the money trickling back begins to flow.
- Along with these "soft factors" EMBLEM also demonstrates the structural challenges of a comprehensive technology transfer process: scouting, patents, administration, and business. It also recommends a technology development fund that lets you push projects in the direction of commercialization and to facilitate start-up projects, an on-site incubator or technology park, and an associated seed fund.

The Spinout Specialists:
Cooperation and Technology Transfer II

Interview by *Samuel R. Schubert*

Great Britain

POSTE
THIS INSCRIPTION IS PLACE
MEMORY OF THE MEMBER
THE MEDICAL SCHOOL
OF THEIR COUNTRY DUF

E TODAY
RPETUATE THE
LLEGE AND OF
THE SERVICE
EARS 1914—1919

SALE TODAY

SECURITY
UNIVERSITY
COLLEGE
LONDON
UNIVE

"Good academics never really trust anybody else because everything is connected to their personal reputation and publishing profile. They never allow other people to do things without checking it 15 times, and when they do finally delegate a task and somebody returns with the results, they question them, and, as I think of it in a business context, become completely incapable of working in a team."

Academics hire management experts. External consultants build an unconventional bridge between university research results and the market. This is an interview about the recipe, about doing away with recipes, and about many years of experience transferring technology through an obstacle-laden intermediate space.

Lodestone Innovation Partners (IP), a small, highly specialized company based in London, has found a way to successfully bridge the divide between the two cultures of academia and industry. It advises universities on how to nurture innovation internally and interact with industry outside. It helps carry ideas from concept to product to market, and from the earliest phases of a project on, it also helps its clients find the investments and management expertise necessary to turn their research into real-life products.

Lodestone IP calls the process with which it increases both the quality and quantity of spinouts "launch management," a method it has used over the past five years to successfully create an impressive niche in the intellectual property market, quickly becoming the paradigm of technology transfer.

One hundred and eighty years ago, UCL (University College London) was the first university in England to admit students of any race, class, or religion, and the first to welcome women on equal terms with men.

UCL was founded in 1826 as the original University of London and currently employs over 5,000 active researchers. Nineteen of the university's academics and graduates have been awarded Nobel Prizes. Today UCL is one of the major centers for future-oriented research and technology in the world.

The college has launched 50 spinout companies, some of which have since listed on a major exchange. Lodestone IP and UCL have worked together on several spinout companies, including Systemwire and Virtual Lightfield, as well as a variety of network development and training initiatives such as "The London Entrepreneurs Challenge" and "Meet the Spin-out."

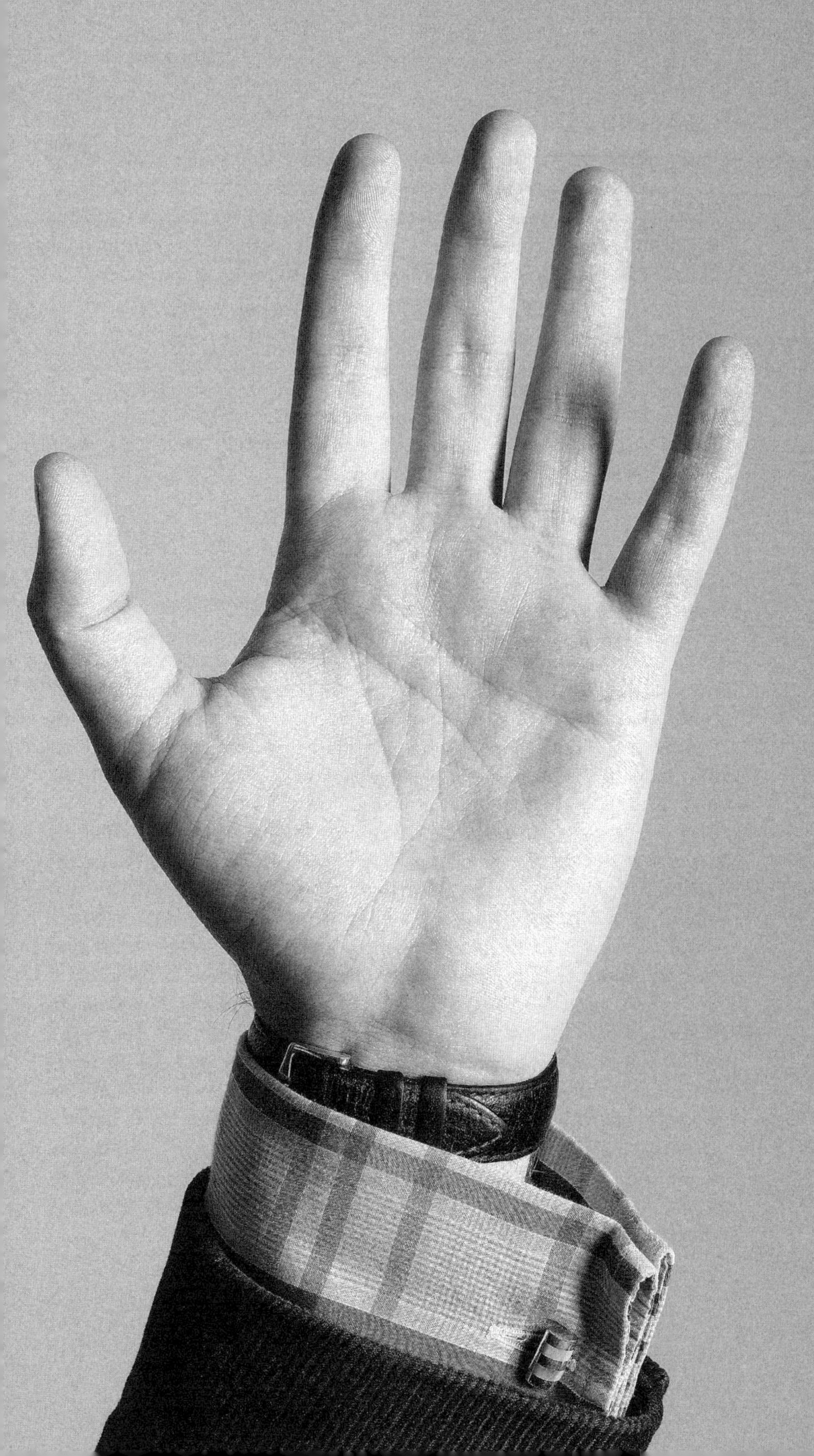

This interview took place in a small corner of the main hall of the university campus. Just beyond its medieval statues, Egyptian stone cats, and the embalmed body of Jeremy Bentham, the university's spiritual father, the three of us met to discuss Lodestone's unique model for producing spinouts and connecting universities to the commercial world. Discussing high-technology university–industry cooperation in such a setting seemed a little surreal at first. The dialogue, however, was anything but.

Timothy Barnes, co-founder of Lodestone IP, a four-person team of consultants based in London. It specializes in providing universities with external help in technology transfer processes.

Peter Sinden, senior technology transfer manager at UCL – University College London.

Ideas and the vision of what could be

Let me begin with a description of two worlds, one academic and one business. I call it a divide. Lodestone and University College London bridged that divide and continue to successfully cross it regularly. In so doing you have gained a unique insight into each other's working environment. Based on those experiences, what conclusions have you drawn, and how would you describe each other's world?

TIMOTHY BARNES/LODESTONE: There's a huge divide, and it comes from the differences in approach. First, there is the issue of trust. Good academics never really trust anybody else because everything is connected to their personal reputation and publishing profile. They never allow other people to do things without checking it 15 times, and when they do finally delegate a task and somebody returns with the results, they question them, and, as I think of it in a business context, become completely incapable of working in a team. In business you have to be able to say: "Okay, these aren't the results I wanted," or "This is a bit surprising, but that's the team's decision, and now we move on." Now, most academics are incredibly bright. They've always been able to overcome problems presented to them. They believe that, given sufficient time, they have the ability to solve any problem. Therefore, they want to solve every problem. And they want it 100 percent right! The idea of being good enough just doesn't exist in academia; it's always got to be perfect. So everything gets revisited again, and again.

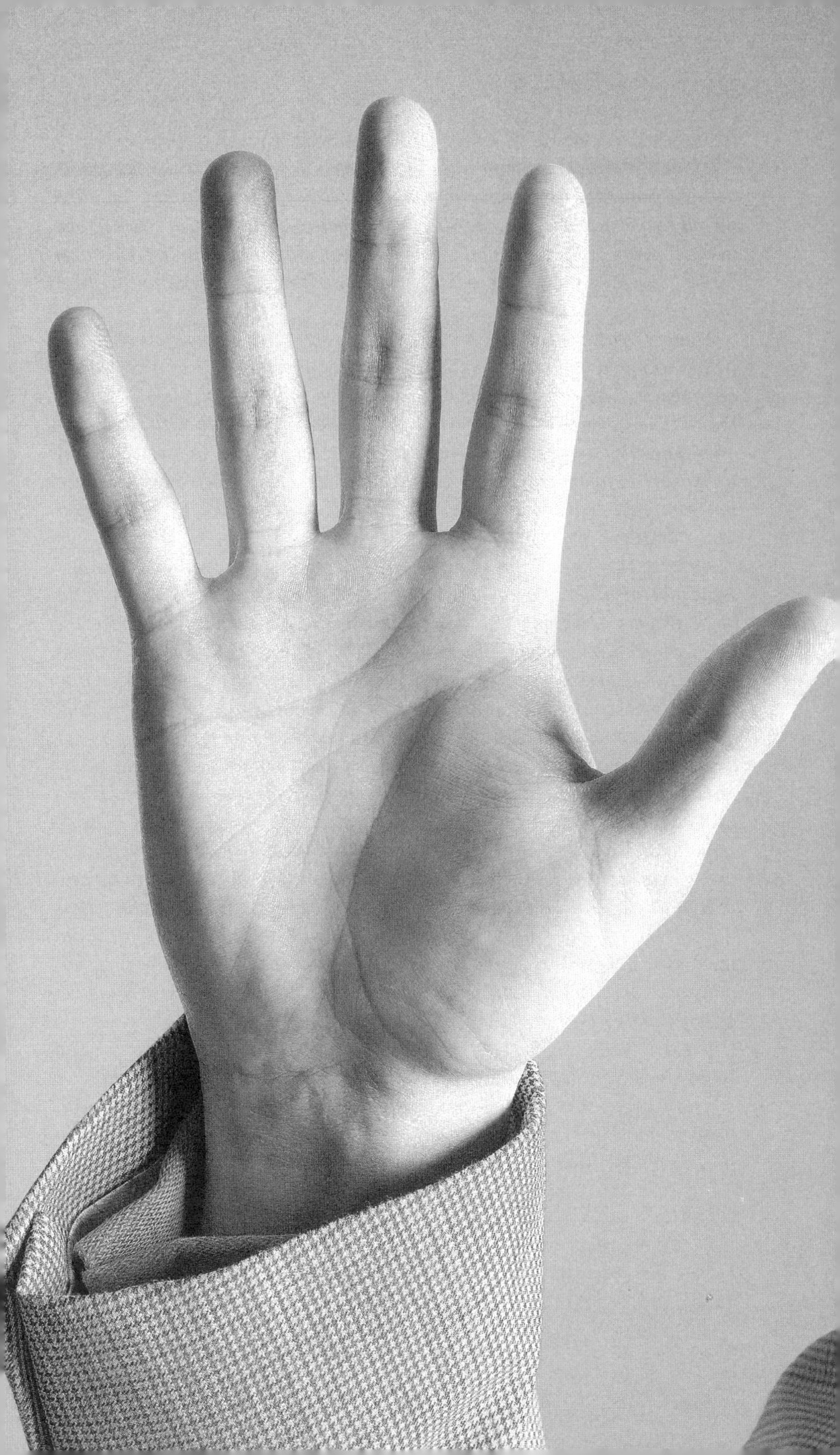

It sounds like you are describing an infinite cycle.

It's not infinite because business people won't sit around while academics debate the merits of something they already decided. They just move on. I find myself frequently repeating the 80:20 rule where the first 20% of effort results in the first 80% of benefit, and the next 80% of effort results in the remaining 20% of benefit. As a businessman, I want most things done well enough rather than perfectly.

Another issue has to do with how problems are solved. A business approach will traditionally entail a specific solution to a problem: one problem equals one answer. The typical academic approach to solving a problem is to look for all the answers. They start with one question and end up with many! It's essentially the reverse pyramid of options. The business I run, that of technology transfer, is all about putting those two pyramids on top of each other, and then navigating through the multiplicity of options to the one point where there is something new.

PETER SINDEN/UCL: I think the two key differences between business and academia are time scales and reward structures. Time scales for business focus on developing something now to sell within a year. If a product has to be on the market three weeks before Christmas, and the scientist needs four more weeks in late November, they are simply going to fail. Academics do not work on that kind of time scale. Their idea is that there are many things in the world that aren't understood, and it will take forever to understand them.

Reward structures also look different. Most academics are not out to get fabulously wealthy; that's just not a driving force for them. Our researchers here at UCL tend to invest over half of their income back into their research rather than taking it home. More important are issues like credibility, respect, and moving up the research ladder, and that can go against the very concept of commercializing an idea in the first place. For academics, protecting ideas through secrecy and patents may limit their ability to publish and constrain their research agendas. The key is showing academics that they can gain credibility and commercialize an idea at the same time by bringing an idea to market as proof of its worth to society.

TIMOTHY BARNES/LODESTONE: Let me add one other point: unless they do something really stupid, decent academics in Europe will get a chair and have a certain amount of research funding and a few PhD students to work with forevermore. They don't have to apply every year for research funding like in the US, and they don't necessarily have to define potential commercial benefits.

Photo   (**see page 105**) Timothy Barnes, co-founder of Lodestone IP

Therefore, it is vastly easier in Europe to do the really big stuff, including ridiculous things like time travel or teleportation. Take, for example, the Human Genome Project. Europeans came up with that. Nowhere else in the world would anybody fund such an exercise.

Mr. Sinden, you are responsible for coordinating these ideas at the university and helping to bring the best ones to the outside world. University College London is one of the biggest research universities in the world with over 5,000 full-time researchers. That is an enormous amount of creative brainpower in one place. How do you keep track of all those great ideas? How do you go about finding, organizing, and evaluating them?

PETER SINDEN/UCL: Commercializing activities out of a university is not about taking every single idea that every academic has ever had and trying to turn it into a widget that someone will buy. That is an approach that a lot of universities have tried in the past. They end up starting 400 spinout companies, all of which fail within six months or as soon as the money has run out. The truth is they were never good ideas in the first place. Now, it would be lovely if every time academics had a marketable idea they would knock on my door and say: How do I do this without messing it all up? Unfortunately, that's not what happens. Ordinarily, I put a lot of work into finding ideas. I knock on the doors of academics and ask how they are coming along. Once we've identified a good idea, then a massive amount of due diligence goes into the process before ever trying to build anything, and it's all about the vision of what could be rather than what is. It's like walking into a really run-down flat, and going: Wow! If we could put investment in it, if we could knock that wall out and develop this bit and put the right curtains in, then this could be a lovely place.

Can you illustrate this diligence?

*Sure. First, ideas are never bad ideas. Second, there are lots of good ideas that people just aren't going to buy. Relaying that to our researchers is essential and there are two steps to that. First, I go out and talk to as many people as possible, find out what is happening within the university, and try to get a really good and global sense for it. **It's just amazing how many academics aren't even aware that they are working on the same problem in similar***

Photo   (**see page 109**) Peter Sinden, Senior Technology Transfer Manager at UCL – University College London

research projects within the university. *I introduce them and return a year later to see what's happened in the meantime. There is a lot of that long-term relationship building at the early stages that hopefully leads to situations later where they'll knock on my door and ask: What can we do next? So one part of it is going out and researching.*

The second step is assessing whether the ideas are commercially viable or not. That means asking what you need to make the idea commercially viable, what kind of investment that would take, and who needs to get behind it. You run that as a thought exercise to see if all the proverbial ducks line up in a row. If it all comes together perfectly, then you have a potential company. But remember, after you've run that thought exercise, you are still stuck with a group of academics with a concept and very little in the way of a physical product or prototype.

You need an idea that makes sense, a market focus, a real belief that it can work, and a strong management team that can realize that vision, one that understands what is needed to put that into practice. When you find the management team to take it forward, that's when you end up with a spinout.

Lodestone Innovation Partners has been very successful in helping universities take ideas and not just bring them to market but turn them into successful enterprises. There must have been a learning curve there. How did you develop your method for doing that?

TIMOTHY BARNES/LODESTONE: Before Lodestone I was with a high-tech investment firm that tried to work with universities. We had five offices across Europe in different countries, and we picked two or three universities in each one and tried to find interesting technologies that we would then help spinout. We tried to treat the university researchers like we treat entrepreneurs. It didn't work. One key point is that venture capitalists tend to approach things in a rather institutionalized, they like to say professionalized, way. They will have a series of tick boxes and fixed assessment criteria. The problem is you're not dealing with the same thing when you're dealing with a university. Academics are not necessarily entrepreneurs. What you do have is a potential solution that has to be shaped and formed. Now, some ideas come ready for market. But some are just proof of concept requiring commercial demonstration. You might have academics who want to join the new company right away and others that don't; some research staff who do, and some who don't; some university staff who are able to assist, and others who are not. In fact, many universities simply don't have the structures to support it. I think one of the biggest problems that companies working with universities have encountered is the idea that there is a rigid formula. We are deliberately set up to work without a rigid formula and always find a way of working with universities.

Is that the strong point of Lodestone?

Possibly. We never approached things saying this is our magic formula for spitting out companies. We approached it saying: Tell us what you've got, we'll tell you if we know anything about it, or if we know somebody who does, and this is what we can do. That means that everything we do with universities is done on a case-by-case basis, and if somebody sends us a project that we know nothing about, or we can't help with, we'll say: Good luck, great idea, we're happy to recommend someone else, if we know them.

Management plus network

PETER SINDEN/UCL: Could I follow up on that? There's essentially a gap between an idea in a lab and a prototype that you can take to a venture capitalist to get funding. This is because of something people call the funding gap. Funding at the earliest stages is the most difficult to get and often the most important. The reason you need funding early is because you need people. It is very rare, particularly in a university setting, that you're buying raw materials, investing in office space, or computing equipment. We've got all of that kind of stuff lying around. But you are, however, investing in the time of an experienced professional who can take an idea designed to solve 50 problems, and decide which one is going to be your first market. You have to identify, pare down, and weigh up the pros or the cons. And it's not necessary to choose the best market, but rather the one in which you've got the best contacts. There are lots of things you need at that stage including someone full time, so you need funding to pay for someone with real experience.

TIMOTHY BARNES/LODESTONE: Yes, not starting with an entrepreneur makes a big difference!

PETER SINDEN/UCL: Right! And it's not just straight management expertise either. You don't want a green MBA student to come and take the idea and put it through a textbook model and see what comes out the other end. MBA students can be useful and valuable resources to take an idea to market, but what you really need at the beginning of a project is someone with a good Rolodex. Connections are the best way to assess real value.

How did you take your experiences as venture capitalists and get to where you are now: offering consulting and counseling to universities?

TIMOTHY BARNES/LODESTONE: My goal was to exploit an opportunity that I thought was there that nobody else was going after. That the universities had a distinct need, as opposed to entrepreneurs and venture capitalists, for advice, help, and support. Nobody was catering to that. What actually happened was that UCL was our first university customer. We walked in to meet with the head of all the technology transfer development activities at UCL and said: We think you've got a problem that no one is helping you with. We want 45 minutes to tell you what we think the problem is and explain what we want to do about it. And we don't want you to interrupt and ask questions because we just want to see if we're right without being moved by whatever clues you might give us along the way. He said okay, and, at the end of it, he sat back in his chair and just said, yes! That was it. We were right on the money with our analysis.

Why UCL?

Very simple. Five years ago, when we began, University College London was one of the best places for doing technology transfer. They had already produced 50 spinout companies, like Arc Therapeutics, which is among the biggest companies to ever come out of a university anywhere in the world. They were also one of the first to set up a dedicated office for technology transfer, and we figured if we could convince them, then we knew we had something.

The reverse question to you, Mr. Sinden, why Lodestone?

PETER SINDEN/UCL: Spinouts start as an idea from a researcher with generally little market focus. I try to convince them to bring it to the real world. Getting that supported by an independent party such as Lodestone adds credibility to that cause. And there are two mechanisms for doing that. One is creating the company around the idea and another is licensing that idea to an outside company. Cooperation with Lodestone is in both areas.

What would you describe then, if you had to pick two or three things as the real strengths of the relationship between UCL and Lodestone?

One of the foundations is the fact that Lodestone, and Tim in particular, understands universities. He's been through a venture capitalist background, but he's not coming at this from the angle of a commercial investor trying to change the way the university works. He also has a good sense for social networking and tends to know a lot of people in a lot of areas or know people who know other people. That is incredibly important. The hardest thing about market research from a university standpoint is that we're researching pretty much everything on the planet. So, having a network to get to people is invaluable. And Tim doesn't work for nothing, but he also doesn't stop the minute that the last ten pounds run out. He has a big-picture stand on things, a good portfolio approach, and I think he realizes the bigger potential.

On pace and loyalty

Mr. Barnes, you had this idea to approach the university and an idea of what you wanted to do. How close did your goals match reality?

TIMOTHY BARNES/LODESTONE: Pretty close with one amazingly huge exception. We approached this project thinking that universities can be quite slow; therefore, we imagined how long it might take us to do a deal and multiplied it by four. We had this wonderful timeline for how long it might take to make a deal and how long before it would work. But I discovered that what I would do in a month in a commercial environment would probably take me a year in a university setting; what takes three to four months in our labs, takes a year to eighteen months in a university. That time lag is the biggest difference I experienced, and it included things like how fast you can build a business as well as the amount of time it takes out of my life.

But there's a flip side to that. Having done an initial sale and built a decent reputation, universities come back and they are incredibly loyal.

How does the university see that?

PETER SINDEN/UCL: I think it works the other way around too. Everybody involved needs to realize that there is a long-term investment in moving these projects forward. Our understanding within the university about how that has to materialize is important because we don't want to spend all our money in the first fortnight and then have nothing for the next three years. It's quite normal for us to call somebody and arrange a meeting, and then only see them again three months later. That is part of the process of moving forward in a university. **The important thing is that ideas that do come out of a university tend to be so far ahead of the market, that you almost need that time to catch up.** *If you get someone to invest 2 million pounds in the first six months and rush your idea from concept to product, the market probably would not understand what you were selling them anyway.*

In addition to the challenges you describe, what was it that astonished you most in a positive way?

For me it is that noticeable ability and willingness of people at all levels to learn, whether within the university or people on the outside like at Lodestone.

TIMOTHY BARNES/LODESTONE: For me that's easy. I walk into universities every day, and I get to see stuff that blows my mind and 20 years from now may completely change the world. My undergraduate degree was in medieval history. Now, I spend my life looking into the future, whether it's in software, engineering, the life sciences, pharmaceuticals, biotech, clean fuel technology, or whatever else, and I get an amazing rush out of that.

Mr. Barnes, you work with other institutions as well. Along the way have you found any specific structures that made cooperation flow easier or observed organizational changes that made a major impact?

I haven't talked to a university in the past five years that didn't reorganize its structure in some way every 18 months to two years. They changed the people working in technology transfer, or combined it with the business development department or research and contract offices, or even changed the way they do licensing and spinouts. The truth is there are just so many ways to skin it. I think the real reason that everyone keeps reorganizing is that nobody yet has hit on the right formula.

So institutional changes are in flux. Are they also necessary?

PETER SINDEN/UCL: They're occurring, but I don't know that they're necessary.

TIMOTHY BARNES/LODESTONE: They're occurring, and it's at the interface part, particularly in the technology transfer arms of the university. And it is happening on the commercial side too. For example, I recently had a conversation with Microsoft about how they want to bring all their research in-house, including all the blue-sky stuff, which used to be the exclusive territory of universities. What that tells me is that there is a grossly inefficient market for research, in which nobody has really come up with an all-convincing case that demonstrates to everyone else that they've found the right way to do it.

PETER SINDEN/UCL: And that comes back to one of my points about the difference in reward structures. If you start a spinout company that gets listed and makes a fortune, that's a very measurable achievement and you've done brilliantly. If you take a lot of research money and produce academic papers and intellectual output, and that's measured scientifically in peer review, then you've done brilliantly. It's everything in the middle that is missing because nobody can really measure it. That element needs to be captured through better forms of technology transfer partnerships.

TIMOTHY BARNES/LODESTONE: And the truth is, the system hasn't evolved yet to cope with that, and I don't know if that'll happen.

From intellectual property to company ownership

The issue of who owns an idea and how it is protected is a vital aspect of any cooperative effort based on intellectual property. Say somebody's got an innovative fuel cell engine design and wants to publish it in *Scientific American* or *Nature*. If one does that, it's gone. How do you calculate and set up the structures to make that work?

PETER SINDEN/UCL: First, you have a very pragmatic thing that starts by making sure your researchers are aware of intellectual property rights and patenting processes before they walk around and say I have this brilliant idea. Believe it or not, that still happens frequently. Now, it doesn't mean that one can't have an article in Nature, *it simply means that we must get the timing right. So if we can talk about it prior to a big conference rather than the week after, then we can get all the paperwork in order before it needs to be done. Universities have to set aside a certain amount of funding to do that.*

Do you have legal counsel in house for handling patent issues?

We at UCL maintain a couple of people in-house with some expertise and have technology transfer managers like myself that have an understanding of what needs to be done. We do work with independent patent attorneys, but we do the basic due diligence before we take it to them in order to minimize their fees. We are not Canon or IBM; we're not filing 500 patents a day. We have a limited budget, and access to great ideas. Technology transfer departments are, therefore, in a sense like a third leg of the university. They advise, encourage, and explain.

When we actually go through the process of patenting, our researchers conduct their own background search, since they know what is out there, and then we take it to the patent attorney when we think it's viable. We can complete the whole process in less than a week if we need to.

So now you've patented the idea, but when it becomes a company, how much does the university or researcher own?

That depends on whether we license the product or build a spinout company. For example, say the idea was a component that would help Canon make better cameras or photocopiers. It would be ridiculous to go out and start your own company to make photocopiers. It is better to license it to Canon so they can improve theirs. There are standard policies within the university following European law whereby the university owns the intellectual property created by those under its employment. Researchers get a proportion of the income derived from that.

The second option is starting a spinout. Its form depends on the involvement of the academic. If our scientist doesn't want involvement in the company, then we'd put together a royalty bearing license, place it into the company, and the academic would get the same royalties as if he or she had licensed it to Canon. We prefer to start spinouts with the active involvement of at least one of the inventors. In that case, there is a balance between royalties and equity, and those can be constructed into myriad different variations.

And how does Lodestone make money?

TIMOTHY BARNES/LODESTONE: First of all, we have a long-term confidentiality agreement with the university that allows us to sit down and talk about projects before any of us have decided whether it's useful or not. That confidentiality is a precursor. Of course, we don't base everything on being paid something in the end. We do little bits along the way and divide them into small chunks.

The first bit might be on a fixed fee, almost like a static consulting project or a monthly retainer. That pays us a little bit, and then we'll do whatever is needed to get the next bit. That way, at any given time we've only got a minimum of ass hanging out of the trouser if the whole thing goes completely wrong.

If an investor comes in, we might take a slice off the top and if at the end of the day the whole thing is successful, another chunk. But, instead of talking about 20 or 30 percent of what's available we might be talking about 5 percent. We nibble all the way along the multiple investment phases until it becomes a workable model without actually killing the entity that we're feeding off.

Those who have tried to do this in the past, whether they be managers or venture capital companies, have tended, for whatever reasons, to use fixed investment models. Those are almost all doomed to fail.

Take an early stage spinout company, for example. If you try to take out too much cash, it'll fall over; it just can't grow. Likewise, if you try to take out too much equity, then the next stage of investor will work you down to nothing and you'll never see any money, or worse, they just won't invest in the first place. What we try to do is blend it all together.

The role of small businesses and rural environments

It looks like your model is something that could work well for small and medium-sized enterprises (SMEs). All spinouts start small and most SMEs need that flexibility in development with which you operate.

PETER SINDEN/UCL: I think there are lessons for universities and companies of any size from all of this. It's important to understand who you're working with and work towards their needs. As a large university, we can be more flexible about the arrangements we negotiate, but like any academic institution, we cannot share the risk on the downside.

We put very stringent policies about warranties and abilities to continue teaching into a license to make sure that we don't put the university at financial risk or our research groups at an academic risk.

Now, as far as negotiating licenses with companies goes, large companies want things for incredibly little, and because they have the power to do that within their supply chain, some believe that they can also do it with intellectual supply, which is not necessarily the case. What they do offer is an ability to accept fairly standard terms with upfront-type arrangements with a fixed royalty over a period, or a straight percentage starting from day one. They're simply less concerned about risk.

Smaller companies, on the other hand, are a lot more concerned about working capital and cash flow management and those kinds of ideas. They are, therefore, much more willing to adopt ideas that are further from the market and put the additional, albeit smaller, investment in themselves. In short, big companies want things they can sell tomorrow. Small companies want things they can use to leapfrog big companies. **Small companies will take a more risky idea, but they'll take it on terms where they essentially pay far less in the short term, but which pay back much more over the long term if successful.**

How does a small company go about engaging a university? Please describe the means by which a small company could cooperate with a university.

There are many ways to engage a university. Anybody, regardless of size, can sponsor a studentship. Buying university time is a cheap business. Now, a small company probably wouldn't have the money for sponsoring a lot of research groups. What they can do, and are far more nimble at, is to pick up ideas out of the university and do something with them. Remember, it's completely free to talk to academics and find out what they are working on. That is important because one of the biggest problems universities have is finding out what the market wants, and that is exactly what small companies are doing all the time. Sometimes small companies have some big advantages. Funding a PhD student for £12,000 or something small like that generally means that the university will own the intellectual property, but the company that funded the research will have something along the lines of a first or exclusive right to license.

Let me give you an example: one of our researchers just stumbled onto something that he'd never expected, something in an entirely different area from the research he was doing. His studies were sponsored over five years by a small company at a time when no one else really wanted to. And because they funded the researcher, they own, at least in part, the intellectual property. When it hits the market, it will completely blow out the scale of that little company; we're talking about something that will increase its revenues a thousand times over.

Modern communications technologies have added a new dimension to university–industry cooperation. The image of the mad professor sitting in his mountain cabin collaborating on the next great scientific breakthrough comes to mind. Is that feasible or is that science fiction in

terms of the real practice of inventing things, bringing them to market, and establishing spinouts?

It's completely workable. We had a PhD student who just recently left the university and is carrying forward with very minimal funding. He already has business offices in other countries, including offshore developments for software in India, and work in Milan, Paris, and Delhi. It is essentially a one-man global operation. Communications technology has been fantastic in putting that forward. There are no global constraints anymore.

TIMOTHY BARNES/LODESTONE: I completely agree with that when it comes to actually building a business. However, with everything before that, I think it's very difficult. Innovation happens in areas of concentration. I had a meeting yesterday with a guy who has a dual research role: working at an academic institution and serving as head of research for one of the largest software companies in the world. We were talking about the fact that great ideas and great innovation comes out of all things multidisciplinary. By that I mean multidisciplinary individuals: people who wake up one day and say: I'm a brilliant scientist, but I recognize that I need others to make my idea function. To do that one needs to be in large communities where there are similar things going on, where you can go and ask somebody questions and talk about plans. I also think it's important that researchers teach and it's very hard to teach virtually from the middle of the highlands. Once you've created an idea, it's another story. It's like writing an essay. You go to a lecture and talk and it's very communicative, but when you're finished you go back to a dark room and sit there and write on your own. You can go off into the highlands, communicate, do a lot of writing, and develop a company separately, but the innovation bit happens in a community.

Community, remote communications, and working face to face

Let's thrash out this community concept a little more. We seem to share a definition of urban as a dense mass of people, but there are new ideas that say urban can be redefined, made independent of physical proximity; that it can be based not on physical, but intellectual mass.

I'm a great believer in face-to-face meetings, where you just go for a coffee and a chat. The reason I think it is so important has to do with push and pull. When you're in the large community you can meet colleagues, discuss and build models, you know the people and you can always get to them. It is that walking around and being stimulated by seeing stuff you've never thought of, that's the push part. It's like the desire of people to still go to the supermarket so that they can see things they haven't thought of and can be presented with a wide range of options at the same time, even stuff they didn't know they wanted. Place that against the online shopping model where you know exactly what you want. You go online, get presented with a very narrow selection, but get the best price. That's the pull part.

They are two entirely different functions. At any given moment one or the other is more appropriate.

Take another example. Go and look at large postal or nonphysical universities, like the Open University in the UK. Compare their technology transfer record with other major institutions, and I'd predict the quantity of truly innovative ideas originating from them is pretty low. I also bet if you go and look at any of the truly huge online universities, big innovation doesn't happen there.

So team innovation requires an urban environment?

I would use the word concentration not urban, but it's putting all of these things in a big jar, shaking it up, and seeing what comes out. It's that random interaction between particles where the excitement and the energy comes from.

PETER SINDEN/UCL: All of the ideas that come out of web-based bloggy, MySpace-type environments don't quite evolve. You could probably cull through online environments in the same way that I troll through colleagues at the university and find something really interesting. But when you do, then you are talking about an entrepreneurial environment where the idea you've found has already been published, so you're going to be at a competitive disadvantage from day one. The truth is that to take an idea and then build a business you need private communities and a personal approach. You need the community to bounce ideas off and you need to be behind a private environment to develop that idea before going out into the public. A physical institution is a very good way of doing that.

TIMOTHY BARNES/LODESTONE: Absolutely! We get together, have a coffee, and chat about what's going to happen in a way in which I wouldn't log on and send an e-mail.

This cooperation and innovation network that you both described seems to involve a lot of personal, even informal interaction.

I've never done an investment deal, not just in celebration, which did not first involve a significant amount of time over a beer.

PETER SINDEN/UCL: I agree. I think informality is important because of a couple of factors. One is the people involved, and the other one is the relationship between them, which needs to be able to be honest and get heated.

TIMOTHY BARNES/LODESTONE: Oh yes! That's a very good point indeed.

PETER SINDEN/UCL: And that's very difficult if either of you can simply log off and not let it really affect your life.

You mean you have to fight?

Yes. You have to be in a position where you can't just walk away.

TIMOTHY BARNES/LODESTONE: That's an incredibly good point. The critical thing is that I need to know that if I tell you that I think you're wrong and, in fact, explode because I had a bad day, that you're going to come back to work tomorrow and still be on this project.

You describe it somewhat like a marriage?

Oh, this is vastly more important. You're going to spend twice as much time with this as you will at home.

Spoken like a true bachelor. Mr. Barnes, Mr. Sinden, thank you for this interview.

Key project insight

- External technology transfer consulting gives a neutral outside view. It supports the university's internal TT management with its specific network and helps academics estimate the market potential of their work.
- Through long-term confidentiality agreements academics are given a protected setting in which to discuss their ideas with market professionals before any decisions regarding their usefulness have been made.
- Important for successful spinout processes are not just capital but management support in the initial start-up stage.
- There are different ways for consultants or capital solicitors to collect their fees: lump sum or payment for services rendered, joint ownership, or profit sharing. Lodestone divides its fees across different stages of every project allowing flexibility, growth, and shared responsibilities.
- Cooperation partners on the industry side need to understand the way academics think and the cultural differences, especially in respect to time scales.
- University management has to expend active effort on technology transfer management. Personal contact, arranging informal encounters, maintaining constant contact with researchers are important for recognizing and developing ideas on the university side.

- Innovation rarely occurs in virtual space. It needs the exchange – including fights – that takes place within a team.
- Long-term research collaboration with individual academics or small teams is a financially transparent model that also has great potential for SMEs.

A New Architecture of Cooperation:
Research Cooperation and Regional Development I

Interview by *Hans-Joachim Gögl*

Austria

B 02

"What is crucial for this sort of model is a really meaningful and ongoing investment in community building, in the development of a culture of cooperation."

Lakeside Science & Technology Park: Company research on a university campus. University institutes and specialized companies under one roof.

Over the past few years a new site of interdisciplinary research and development, education and training, production and service has been created in the Austrian federal state of Carinthia. The facility, which when completed will cover an area of 28,000 m^2, currently employs 500 people and is located close to the University of Klagenfurt.

Companies and research institutes whose main fields of focus are selected topics in information and communication technology gather here to collaborate on R&D projects: the technology park as an ongoing workshop of company development and university research with experts from the sectors of business, technology, and the cultural sciences.

An interview about research cooperation as a development impetus for the rural area, chances and challenges of a "peaceful coexistence" between company and university researchers, and the game rules and experiences of building a site for innovation out in the sticks, or as it were on the banks of Lake Wörthersee.

Since 1998, Erhard Juritsch, Mag., along with Hans Schönegger, Mag., have been on the Board of Directors of the KWF – Carinthian Economic Promotion Fund – which on behalf of the Carinthian state government oversees economic funding in the region. Both are the initiators and managing directors of the Lakeside Science & Technology Park. In addition to working to promote economic and regional development, Erhard Juritsch is also currently writing his dissertation on entrepreneurial decisions concerning investments in foreign regions.

University Professor Dr. Heinrich C. Mayr, one of Lakeside Park's founding fathers, is currently the president of the University of Klagenfurt, having been previously the dean of the Faculty of Business and Computer Science. He studied in Karlsruhe, earned his PhD in Grenoble, ran a software company in Germany, and has been a full professor at the University of Klagenfurt since 1990.

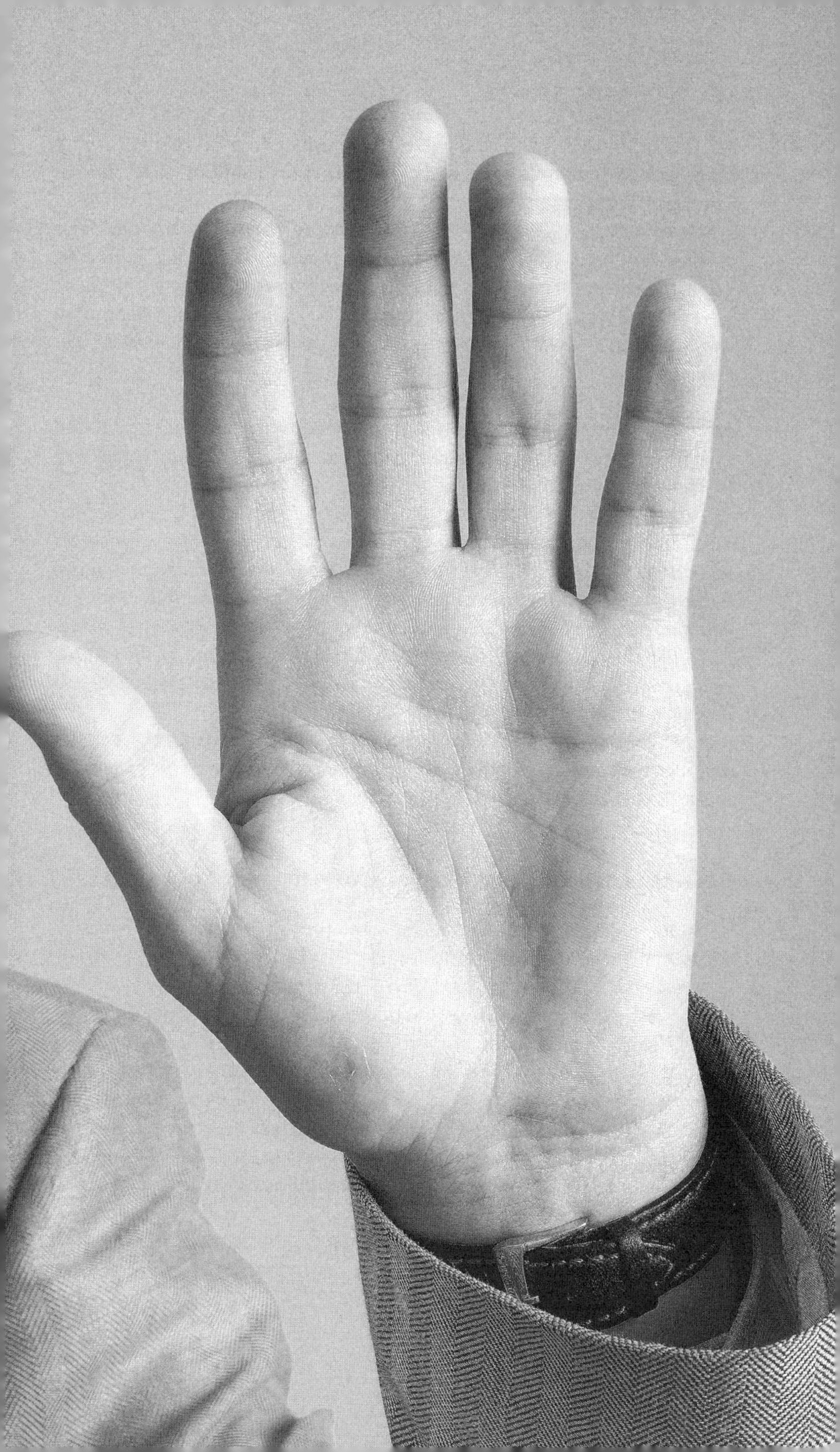

Mr. Juritsch, supposing you were the president of the University of Klagenfurt, what would you wish for from your neighbor, the Lakeside Science & Technology Park?

ERHARD JURITSCH/LAKESIDE PARK: A clear course of cooperation in a wide range of directions: right from the start tenant companies at Lakeside Park offer students a glimpse of normal working life in their future professions, a valuable asset that could set us apart from other universities. Then, intensive cooperation on research and development projects and support in financing matters, e.g. public/private.

Professor Mayr, in this role-swapping scenario, I give you the reverse question: as the managing director of Lakeside Park, what would you wish for from the university?

HEINRICH C. MAYR/UNIVERSITY OF KLAGENFURT: Lakeside Park is a regional development project whose aim is to support existing technology companies and attract new ones, and that can't work without the right human resources. Particularly with a business structure like you have in Carinthia, which is dominated by small and medium-sized companies, you need the kind of support offered by an external research facility such as the university because small or medium-sized enterprises, SMEs, can't afford to conduct research and development on their own.

Having said that, I'd wish for practically oriented training for the students and research cooperation projects tailored to meet the needs of the companies in this economic area.

Research cooperation as a regional development strategy

As to the regional political strategy of the project, what were the overall goals that led to the decision to implement this type of model?

ERHARD JURITSCH/LAKESIDE PARK: Since 1945, economic promotion in Austria, and probably throughout Europe, has traditionally been divided out indiscriminately, supporting as many recipients as possible with the least amount of funding. With this project we took a different approach, we allotted a large amount of funding, admittedly as an investment, with the expectation that the Park will pay itself off in the medium term. This project focuses on two main strategies: first, the emphasis on future-oriented technologies, and second, bundling. Carinthia continues to lose jobs in classic industrial areas and trades every year...

HEINRICH C. MAYR/UNIVERSITY OF KLAGENFURT: ... in fact, the second to last shoe factory in the state just closed its doors yesterday!

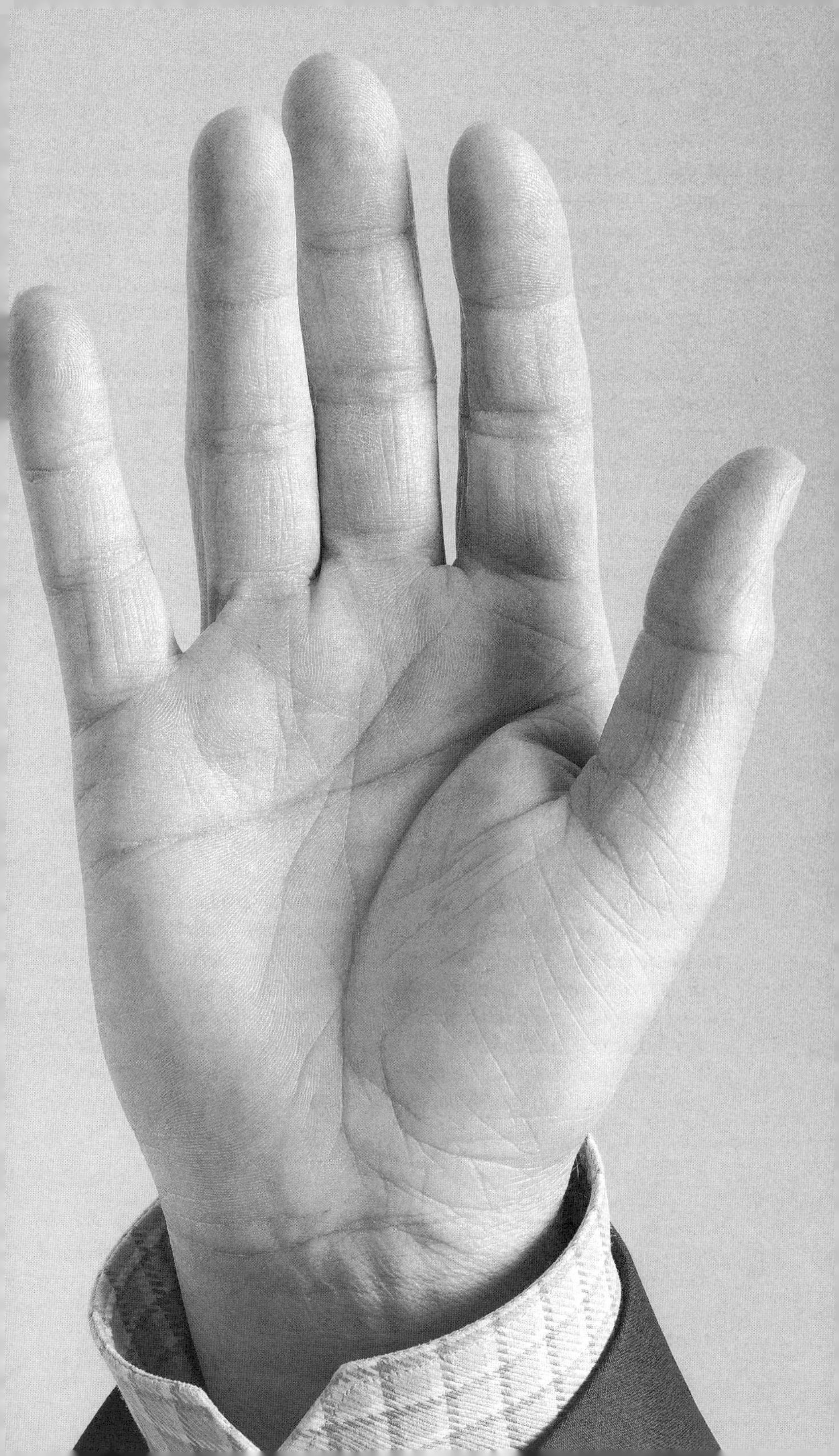

ERHARD JURITSCH/LAKESIDE PARK: ...Exactly, and in order to compensate for this trend in the long term, we came up with Lakeside Park as an initiative to develop sustainable occupational fields in the region with growth potential for the future. Not that we think every valley should have its own small-scale project. On the contrary, the idea is to bundle all the regional energy and resources at a given site and offer something extraordinary in combination with the university and other leading companies in the relevant sectors.

For the companies themselves there is another highly innovative strategic aspect about the Park, aside from the strong potential for university collaboration: today we are witnessing a twofold trend in which companies are moving toward extreme specialization, while clients are looking for increasingly comprehensive system solutions. Companies which are capable of organizing themselves like a fleet to flexibly fulfill individualized customer wishes can offer both. To this end the Park is a strategic platform from an entrepreneurial perspective as well.

You can't simply set this type of facility in the middle of nowhere and prescribe inspiration to the region. You need an economic area where you can get real resonance sounding off structures that have developed over time. Is this the case in Carinthia?

Yes, and that is the crucial point, not everything is possible everywhere just because you have the financial means. In Carinthia we have an interesting structure of IT companies, large ones like Infineon, quite a few high-tech SMEs, plus an internationally successful computer science department. In other words, we are strengthening an already existing strength.

In addition to the effects generated internally, the Park also serves as a future vision for the region. It shows what work models will look like in the future. *In order to communicate this, we need not only concepts but real role models because the Park was not intended merely as property for lease; rather, it should serve as an inspiration and motivational force, an example for other initiatives. In the planning phase we took our political decision makers and other stakeholders around to the best technology parks in Europe to show them in as concrete a way as possible how successful projects work.*

Photo   (**see page 139**) Erhard Juritsch, co-initiator and co-Managing Director of the Lakeside Science & Technology Park

Professor Mayr, what possibilities are open to a technology-oriented university located in a rural area? Klagenfurt has no metropolises to support it and consequently no large pool of potential students nearby to draw from. Your department of computer science is smaller than all its German counterparts, and yet university and technology park postulate international excellence and are building an IT center whose impact extends far beyond their regional borders.

HEINRICH C. MAYR/UNIVERSITY OF KLAGENFURT: The key here is a two-pronged approach: specialization and applied research. From the beginning we placed great store on collaboration with local companies and developed an applied computer science department that focused on practice-oriented basic research and has worked with companies from the start. This is reflected on two levels: I believe we were the first school in the German-speaking world to integrate internships into our curriculum; back then it was still quite uncommon. And we not only received the usual standard third-party funds, e.g. from the federal government or the EU, but from the beginning we have also consistently conducted research projects with industry.

In the most important academic ranking of all German-speaking computer science universities (source: CHE – Center for Higher Education Development, Gütersloh) we are number 1 in 12 of 20 categories. Overall, we are among the top 5 out of 82 computer science study programs in Germany, Austria, and Switzerland. By virtue of its size, the University of Klagenfurt may not be able to really stir things up, but in the fields we teach and do research in, we have an outstanding international reputation.

This practice-oriented focus in computer science has meanwhile spread to the entire university. **We see ourselves as a regional university that takes its responsibility for this area seriously, that follows a consistent niche specialization strategy in key fields, and which for its size offers a wide academic range.**

In order to attract students from outside of Carinthia, you have to be able to offer something special. And that brings us to Lakeside Park, which lets us collaborate in sophisticated applied research and offers interesting jobs. We're banking on a kind of spiral: if we can convince a lot of attractive industry partners to join us, this will get more students to come to us, and that will encourage more companies to set up shop here, and so on and so forth.

Photo   (**see page 143**) Heinrich C. Mayr, President of the University of Klagenfurt

Mr. Juritsch, what has been your experience in respect to technological specialization in the periphery? One of the things Lakeside Park specializes in is the field of transportation telematics. To what extent are globally active companies willing to shift research units from their headquarters to southern Austria near the border with Slovenia and Italy?

ERHARD JURITSCH/LAKESIDE PARK: The idea is no longer to spread entire units around the globe, but to participate in strategically important projects. For example a large corporation might only send one expert here. Through this expert it would systematically assign thesis subjects, and when the project took off, it would have outgrown the existing surroundings. To us there is an aspect of sustainability involved too. That means that what counts with a potential tenant is not the size of the department it sends, but the growth potential of the given project. This is also how we determine the success of the Park: what is the potential of each project?

We're not interested in renting space to stagnating companies. So much about our strategy for large companies.

Small enterprises come here because they can expect to find a nurturing environment with the university, venture capital, good counseling, and an interesting network.

How do you build a site for innovation?

During the conception and construction of the Park, what were the main considerations for creating a site that would be as conducive to cooperation and innovation as possible?

*Along with a conscious selection of the tenants, our aim was to create informal, everyday spaces of encounter between the players. There are several university institutes located directly on Park grounds, students and professors work among the companies, there is a restaurant, a café, the Kunstraum art space, a kindergarten – all of which may, on a short-term level, seem almost unintentional, but which are in fact in the overall scheme very conscious elements conceived as places to encourage informal contact between the companies and the university. In addition, there is the high architectural quality of the site and of course its beautiful location here on the banks of Lake Wörthersee. **Promoting cooperation isn't first and foremost about being goal-oriented, but about opening up spaces in a playful way.***

Which structures and instruments has the university created to encourage and support contact with companies?

HEINRICH C. MAYR/UNIVERSITY OF KLAGENFURT: We set up a research and development aid office for colleagues who lack experience in cooperation projects. The office helps them fill out applications, explains basic legal or economic structures, etc. We have our own lecture forum to which the companies are invited regularly, and we also founded the Lakeside Synergy Club. It consists of organized, unstructured gatherings, in which we present ourselves and our work to the companies at Lakeside Park, and vice versa. To date, our only investment has been coffee and cookies, but it has resulted in a number of new cooperation projects between the university and individual companies.

ERHARD JURITSCH/LAKESIDE PARK: The Synergy Club is a good example for illustrating what works. It's all about processes, and our job is to create platforms and encounters that allow these processes to take place.

HEINRICH C. MAYR/UNIVERSITY OF KLAGENFURT: All you really need is a big coffee-vending machine where people can congregate. This kind of situation often leads to innovation, especially in technical fields.

On the economy of teaching and research

Aside from joint research and development, how important is recruiting in getting companies to settle at the Park?

ERHARD JURITSCH/LAKESIDE PARK: Very. There is intense competition for the best minds, and as a company, the closer I am to the university, and that includes physical proximity, the easier it is for me to build my network.

A park like this one is an institutionalized process that allows us to reconcile the differences between the expectations of educational demand on the part of industry and those of educational supply on the part of the university. The quality of this dialogue will help to determine overall success. In Europe the birth rate is decreasing and so is the interest in studying technology subjects. This means our existing potential must be used as efficiently as possible. And that is why it is important for companies to communicate as precisely as possible which skills are needed today, and, as far as they can predict, which ones will be necessary in the near future. At a place like Lakeside Park, located directly on the university campus, this communication happens almost automatically in the sense that professors get direct feedback from managers and interns, and can react quickly to adjust the supply.

How does this sound to the ears of a university president? Are you committed to the principles of open teaching and basic research? With this you-scratch-my-back-I'll-scratch-yours model don't you run the risk of reducing the university to a training center driven by purely economic interests?

HEINRICH C. MAYR/UNIVERSITY OF KLAGENFURT: No, companies don't place orders with us like: we want you to train ten computer science experts with these skills and those qualifications. Besides, it's something we very well might do, but only as part of an advanced training program. The education we offer is designed to be broad enough – as opposed to technical colleges, which operate in a more job-oriented way – that our students will be able to learn what their bosses need in a very short time. We think this is a better approach because in the long term this kind of education makes you much more versatile, as opposed to specialists with an expiration date.

Another one of our principles is that we won't agree to any cooperation projects unless they also serve academic goals, e.g. the publication of research results. Furthermore, our partners know that we expect to be paid reasonably for solutions to applied problems, that the university is not prepared to provide company sponsoring.

But if an industry partner is interested in a specific international instructor who teaches material we can't offer, and provided it fits into our curriculum, we would be pragmatic about it and invite him or her to lecture to our students, and company staff are welcome to come and listen. Or if a company approaches us and asks if we can jointly analyze a certain problem, we have been known to hold seminars in which we took their actual case studies and worked through them with the students. Everybody profits from these practical examples and the feedback from the company.

The diverging views of companies and universities have evolved from different mentalities and historical traditions. How well are these differing organizational cultures equipped to engage in collaboration today?

ERHARD JURITSCH/LAKESIDE PARK: These differences do exist. The focus in the company is on the quality of the economic outcome, on quick market success, while I would say that what is most important to universities is knowledge. That means, among other things, that there is little correlation. But this difference can be an advantage too, and one always profits from dealing with others. One group might learn about pragmatism and project structure, the other about thoroughness, depth, and diversity of perspectives. Incidentally, companies can be slow too, and the larger the company, the slower it becomes. I know corporations where it takes longer to get a signature than going through the red tape of a government agency.

HEINRICH C. MAYR/UNIVERSITY OF KLAGENFURT: Austrian universities have made a lot of progress in recent years. Today, we are structurally more like a company with a very short decision-making process. The president is more or less the CEO – what he signs goes into effect. If we want to close a normal deal – a project as opposed to alterations to the charter, for example, which would entail a committee decision and could drag on for years – we can take care of it in three days!

Before you got involved with the university, you ran a software company for several years, so you've experienced cooperation from both sides. What advice would you give a colleague who was getting ready to start a cooperation project with a company for the first time?

It's important to clearly understand the company's expectations. For this you must invest time even before you have actually decided to work together. After that, you assess your own time and the human resources you will need, which will all have to be mobilized parallel to your teaching activities. And what comes out in the end of this kind of cooperation project isn't a research report like you write for EU projects but a practicable solution for the company. And for a scientist who hasn't had much experience with companies this is a whole new ball game!

In your experience, what are the most common mistakes in cooperation projects between universities and companies?

From the academic side, that task and effort tend to be underestimated, and from the other side, that the company generally has little patience in regard to implementation. Some companies even expect results to have a successful impact in the next quarterly report. We, on the other hand, are interested in the innovation itself, something that can be published and will help us make scientific progress.

Not discussing these disparate expectations is probably the most commonly made mistake when it comes to collaboration.

ERHARD JURITSCH/LAKESIDE PARK: I think this process is precisely where we have to invest time and energy, even though it doesn't directly produce anything. And I also recommend consulting external project management experts. They are not involved in the cooperation project and can therefore remain neutral.

Cooperation projects demand skills that researchers normally don't learn in the course of their studies. To what extent do competencies like facilitating, steering projects, giving presentations, etc. become an issue for students in the fields of technology or natural science?

HEINRICH C. MAYR/UNIVERSITY OF KLAGENFURT: Ah, but we do teach these additional skills. Our approach differs from that of humanities, which tend to focus on the individual expert; by contrast, we concentrate on teamwork. Even in the first semester we don't accept assignments turned in by individuals; students have to work together, from day one all the way to graduation.

Focus, selection, and reserving funds for working dinners

To what extent do the companies at the Park see themselves as a team? Does the Park management specifically encourage project-related collaboration? Or put it this way: would a large-scale software engineering system solution involving many Lakeside tenants be a conceivable vision?

ERHARD JURITSCH/LAKESIDE PARK: A vision, yes, but the main aspect is still our systematic selection of tenants. We turn applicants away regularly if they don't fit in perfectly with the Park's strategically technological orientation. We also know which areas of expertise still need to be filled. We are aware of the companies our tenants would most like to work with, and we specifically approach them. Having these things in common automatically makes it easier to connect. **The worst mistake a technology park can make is to have an arbitrary mix of companies, to accept tenants just to collect the rent and be able to amortize the investment costs quickly.**

We always keep space available to be able to react flexibly to new tenants or to allow expansion if there is an internal demand for it. In addition, we also conduct preliminary tests in which we check the economic stability of the companies, which means here you can rely on the creditworthiness of your potential cooperation partners.

The Park management makes sure that everyone knows what the other tenants are working on. Company profiles are constantly updated, and regular information events like the Synergy Club are held to keep information about each member up to date.

Professor Mayr, just before you settled at the Park in 2004 you said in an interview: "By 2010, if someone brings up the subject of European IT research and development centers, I would expect the name Lakeside Park in Klagenfurt to come up at least right behind Sofia Antipolis [technology park near Cannes]. We want to become a major, internationally relevant center, specifically focused on computer science and IT." How do you feel about your statement today?

HEINRICH C. MAYR/UNIVERSITY OF KLAGENFURT: I would correct it slightly: it's all still true except the year has to be pushed back for the following reason: as Erhard Juritsch has mentioned, we are currently experiencing a steep drop in students in all technology fields. This is a pan-European trend, and reciprocally it is exactly these subjects that are booming at Asian universities. I and many others observe this with great concern. We have to take political action; the universities won't be able to handle this situation alone.

But I'm also optimistic because we have excellent people applying for our newly created professorships. We are regarded as a place to be taken seriously. Along with the good rankings, we are also well represented in the publication indexes. Moreover, during the past few years Klagenfurt has probably hosted more international, high-level computer science conventions than any other university, despite the fact that we are small compared to Kaiserslautern, Karlsruhe, Vienna, or Zürich. In our fields of focus we are developing into a small but first-rate site on the research map.

Mr. Juritsch, based on your experiences of the past two years, what in your view are the most important points for other regions to keep in mind if they want to implement a similar project?

ERHARD JURITSCH/LAKESIDE PARK: First, to carefully analyze the focus of the park and to define this aspect in conjunction with all the stakeholders because their participation is decisive for the success of the project.

And under no circumstances should you invest everything you have in your facilities. I recommend reserving funds for cooperation projects as well. Even if they are just working dinners. What is crucial for this sort of model is a really meaningful and ongoing investment in community building, in the development of a culture of cooperation.

Thank you for the interview.

Key project insight

- A careful analysis of the existing regional competencies.
- Integration of the stakeholders into all aspects of the project: politics, economic promotion, university, technical colleges, and economy.
- Consistent focus on a few specialty fields.
- The goal is to become a global leader in this niche.
- Strict selection of the tenants at the Park and their synergetic coordination with each other.
- Setting up the facility on the university campus and/or integration of institute buildings on facility grounds.
- Supporting the university with additional endowed professorships. In 2007 the Faculty of Technical Sciences was introduced to the University of Klagenfurt. Its main departments include Informatics, Information and Communication Technology, and Technical Mathematics. This was made possible by six endowed professorships: Mobile

Systems, Transportation Informatics, Media Engineering, Pervasive Computing, Embedded Systems, and Applied Mechatronics.
- Staff and budget resources for active community building (cooperation-promoting events).
- Creation of spaces for informal encounters between university and company researchers at the Park: café, restaurant, art space, kinder-garten...
- Sense of beauty as the expression of esteem for the project and its participants: site and architecture.

The Rolls-Royce of Cooperation Models:
Universities as an Integral Part of the R&D Units of a Company

Interview by *Samuel R. Schubert*

Great Britain

10
STOP
ALL VISITORS MUST REPORT
TO SECURITY RECEPTION
ALL PASSES MUST BE SHOWN
VEHICLES MAY BE SEARCHED
CCTV SURVEILLANCE
CAMERA'S PROHIBITED
NO SMOKING
PEDESTRIAN
ACCESS
ALL PASSES
MUST BE SHOWN

Rolls-Royce
2
Visitors must report to reception
Identity passes must be shown
Vehicles may be searched
CCTV Surveillance
Goods Vehicles
Speed limit 10mph
Cameras prohibited
No Smoking

ROLLS
ROYCE
RR
Rolls-Roy

"I think the one thing that you can learn from the Rolls-Royce UTC model is that it's about the time that the company is putting in. Giving us a contract and then turning up every three months and saying: Where's my stuff? just isn't going to work. It's all about keeping people in the university as part of the team, keeping some involvement, and, actually it sounds strange, but growing a lot of loyalty to that company. And that doesn't come easily. When it is there, that is when researchers will go the extra mile."

Rolls-Royce probably practices one of the most intense forms of cooperation between companies and universities. Over the past 15 years Rolls-Royce has completely outsourced vast parts of its basic scientific research divisions to universities. At 26 so-called University Technology Centres (UTCs), academic and commercial researchers work together closely on university campuses around the globe. Unique situation or trend?

Rolls-Royce's motto is "trusted to deliver excellence," and indeed, it is excellence they provide. No matter what they do, whether in the field of energy, marine technology, or aerospace, Rolls-Royce's products are considered the most reliable on the market.

The fact is, if you have ever traveled by commercial jetliner, the chances are it was their engines that got you there. One would imagine such a company having thousands of scientists conducting advanced basic research under the secure wraps of secluded laboratories. However, that is not the way it is done.

Rolls-Royce has long since expanded its research activities from strictly in-house product development run by its own engineers to universities around the world. But Rolls-Royce didn't just stop at sponsoring studentships and putting up external funding for research teams and their designated projects. It has set up an intellectual network that links all these different academic institutes around the globe with Rolls-Royce, allowing the open exchange of theory and practice, of academic and commercial expertise. It's more than just a marriage of convenience.

While the UTCs give Rolls-Royce access to the ideas of the world's leading engineering talents, the universities enjoy the luxury of conducting practical research in stable, well-financed, and commercial environments.

Dr. David Clarke, head of technology strategy and research at Rolls-Royce.

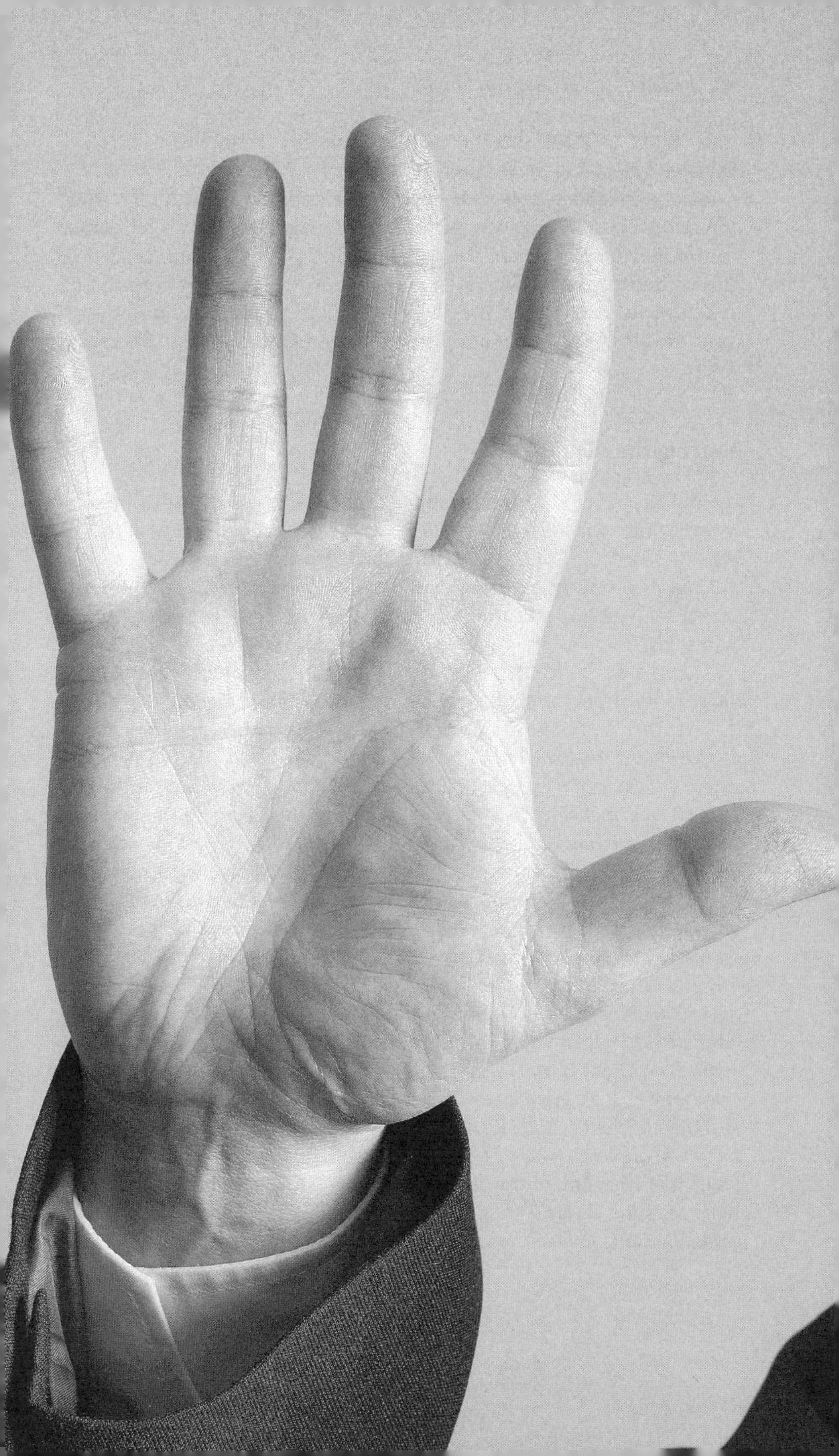

Dr. Geraint Jewell, director of the University of Sheffield's UTC for Advanced Electrical Machines and Drives.

On a cold and wet British morning, I had the opportunity to sit down with two key players of the Rolls-Royce University Technology Centre network, one from the company and the other from academia. The conference room where we met was at the company's major site in Derby, in the heart of the British Midlands, some 20 minutes drive from Nottingham and Robin Hood's legendary home of Sherwood Forest.

Blurring the boundaries

I would like to begin with a metaphor. Many describe university–industry cooperation with different types of imagery. Some speak of different worlds, others of rivers, even oceans apart. No matter how one describes it, there is certainly a cultural gap between approaches, organizational structures, working cultures, and goals. Do you also see it as a divide to be bridged?

GERAINT JEWELL/SHEFFIELD UNIVERSITY: Universities have always worked on a much looser approach to research than industry. Understanding that is critical. It's not the university's job to ape an industrial organization but to provide that distinctive flavor of a longer-term view, the harnessing of enthusiasm, particularly of people arguably in their prime, from a creative point of view, and matching that up with a company and blurring the interface. Obviously, universities have to engage with more formal project management and structures than they would if they were left to their own devices. I do think a lot of companies work with universities to access that different culture, and it works because of that. But it's definitely something that needs to be bridged.

DAVID CLARKE/ROLLS-ROYCE: We at Rolls-Royce recognized that divide and made a deliberate policy of trying to blur those boundaries. Academic research has historically been very curiosity driven. Industry, on the other hand, is very applied, and there is a whole different mind-set that goes with that.

When we looked around most university groups, we at Rolls-Royce tended to find that different culture. But I think that if you look around the world today, those two are coming closer together. What we start to see now is a complete blurring of boundaries between companies and universities in every aspect. In some cases this divide does still exist, but if you get people to work together, it gradually starts to break down.

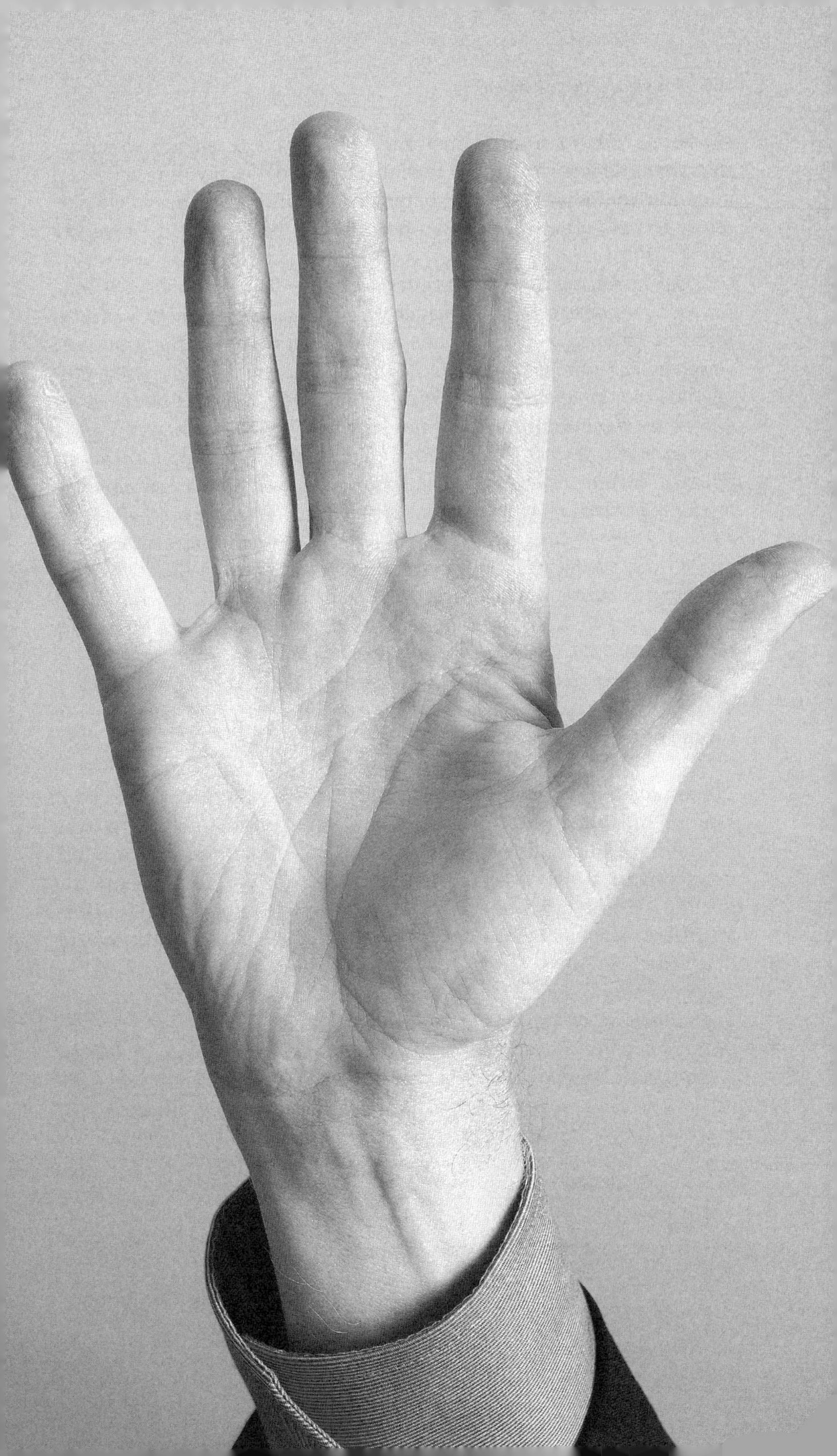

Dr. Jewell, You mentioned that universities have to engage structures that are more formal in order to blur the boundaries dividing academia and industry. What kinds of structural or operational changes were necessary to make that cooperation flourish between Sheffield University and Rolls-Royce?

GERAINT JEWELL/SHEFFIELD UNIVERSITY: Discipline really makes the difference. University research, the time frames on which it's monitored and the outcomes that it is historically judged by, are very different to an R&D program in a company. For example, we might be used to continuously supervising PhD students with an appreciation of their general progress, but the outcomes are only drawn together infrequently when papers or the dissertations are drafted. In other words, it's not really driven by time scales. Working with industry, however, you have to get that balance right: you need regular review meetings, regular milestones, much more project planning, and you have to get all of that without destroying the freedom that you get in the university environment. That's really the big challenge, getting that right balance of discipline versus freedom and harnessing that nicely with the understandable demands of an industrial partner. Now, that's quite a difficult thing to do! Overall, I can say the guys in our place found the cooperation with Rolls-Royce very beneficial to their research in terms of context, encouragement, and constraints; all those things actually make research a little bit more interesting than plowing on in isolation.

Interesting, but isn't discipline in a university something akin to an oxymoron? In companies, it fits. There are hierarchical structures, a clear chain-of-command, and discipline is central in planning and product development. Universities, on the other hand, thrive on all the opposite qualities. Are the types of cultural changes you described possible across an entire academic institution, or are they limited to specific teams or departments?

I think trying to institutionalize any kind of culture in a university is virtually impossible. Do you know the old phrase about trying to herd cats? Across an institution, university academics are a disparate, occasionally bloody-minded body of people. Within a group that engages closely, like a pocket of four or five academics, where each understands the requirement of why impositions are put in place for understandable reasons, you can get that, but institution-wide, no.

Photo (**see page 167**) Dr. David Clarke, head of Technology Strategy and Research at Rolls-Royce

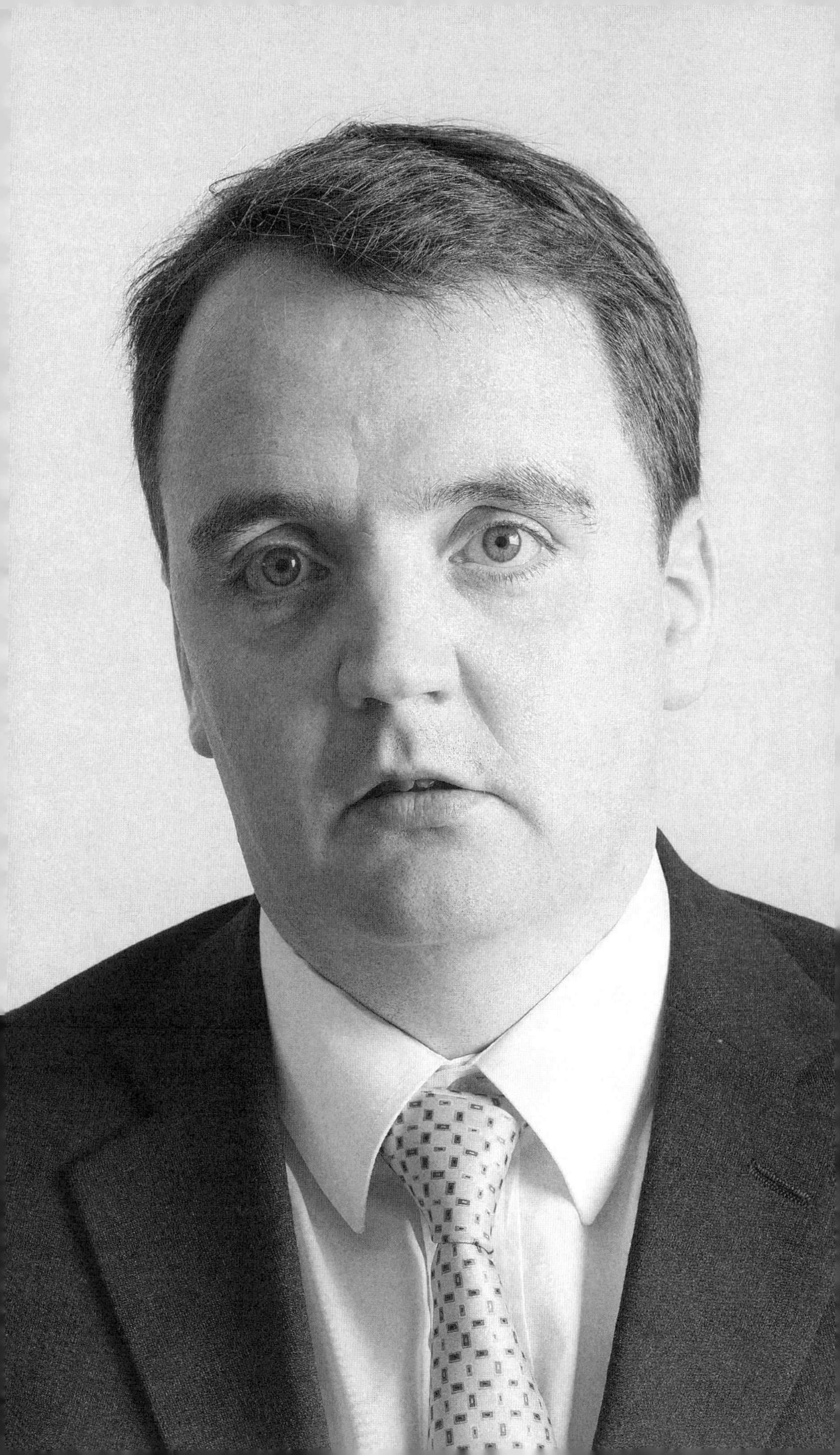

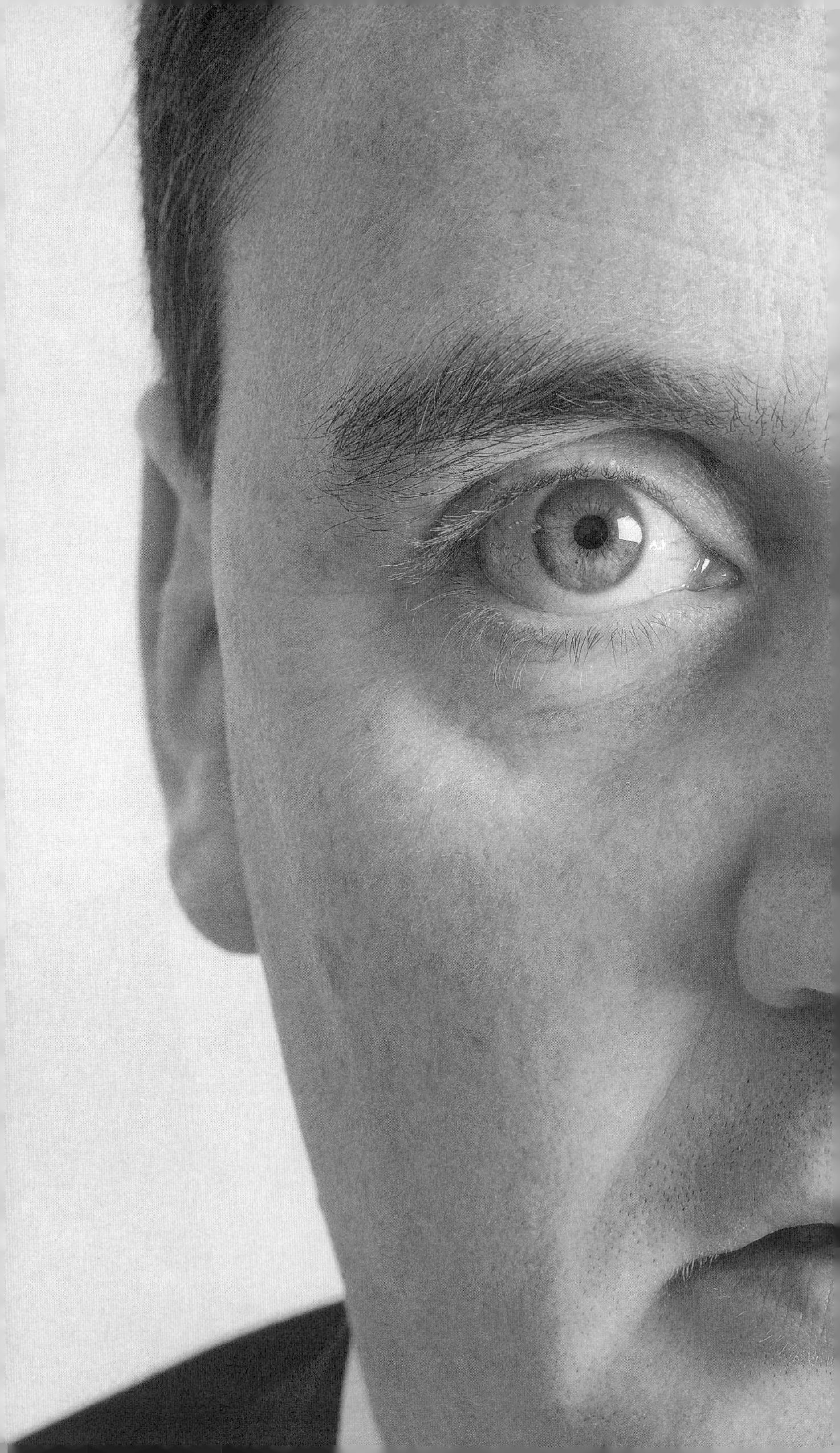

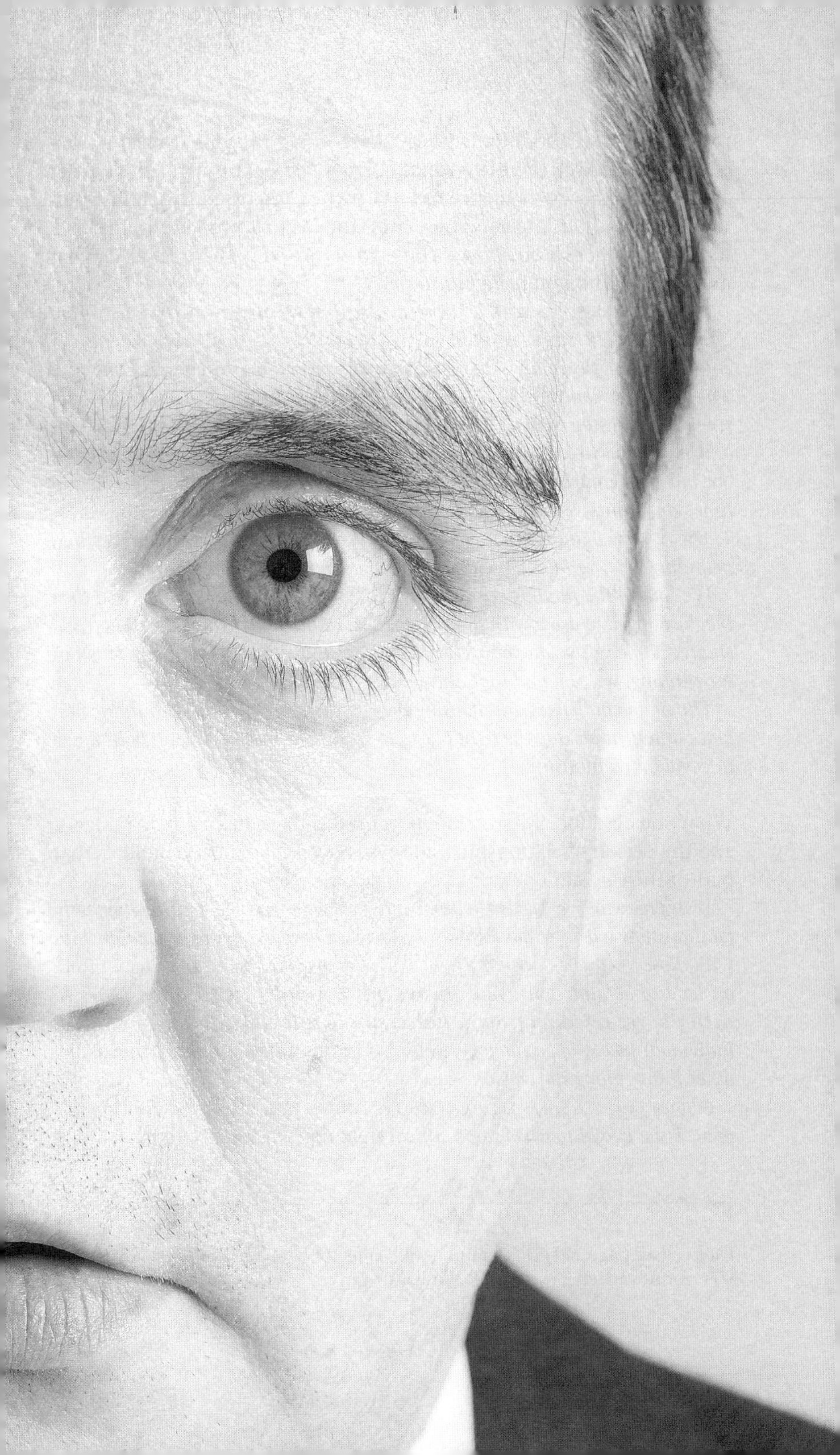

Dr. Clarke, what about Rolls-Royce? One would imagine from the company's perspective there are unique issues in dealing with universities and some of those elements have been just discussed: discipline versus flexibility, herding cats, timelines, intellectual property rights, and so on. How does Rolls-Royce come to terms with these issues? What mechanisms did you have put in place?

DAVID CLARKE/ROLLS-ROYCE: We recognize that we take away flexibility from universities. Therefore, we had to give some too. It's really been a case of enforcing that flexibility. We don't manage everything to one deadline. We don't say, well, once we've started a program that we've now got flexibility in six months to stop it. Once Rolls-Royce commits long-term support to students and postdocs, we continue that and it does require a degree of rigor. Now, our business groups are obviously under pressure to deliver against budget and time constraints, which in some cases goes against doing long-term research. The way we've handled that is by controlling our interaction with universities, including who we interact with and the broad guidelines of how.

We control the funding centrally, allocate it to our business groups and then they're obliged to use it with the universities, rather than being free to chop and change. I mean, we do change things, a lot! But actually, once relationships are running, we've found that universities can be quite flexible.

The main challenges are choosing the right partner to work with in the first place and then picking the right places to work. The reality is, it all boils down to personal relationships.

What you describe, the rigor, support for postdocs, flexibility in budgets, and the personal relationships, all these things take time to build. Is that part of the calculation?

*Yes! The way we tackled it from day one was that we wanted long-term relationships with the universities. In fact, we needed long-term relationships with those researchers with the right skills because the time scales that we work on tend to be quite long, **not always but normally eight to ten years to actually get research from a university to a product.** It's not just about individual pieces of technology; you also have to integrate those with many other things along the way.*

GERAINT JEWELL/SHEFFIELD UNIVERSITY: I think that the long-term commitment issue David touched on is critical from the university point of view and

---

Photo   (**see page 171**) Dr. Geraint Jewell, Director of The University of Sheffield's UTC for Advanced Electrical Machines and Drives

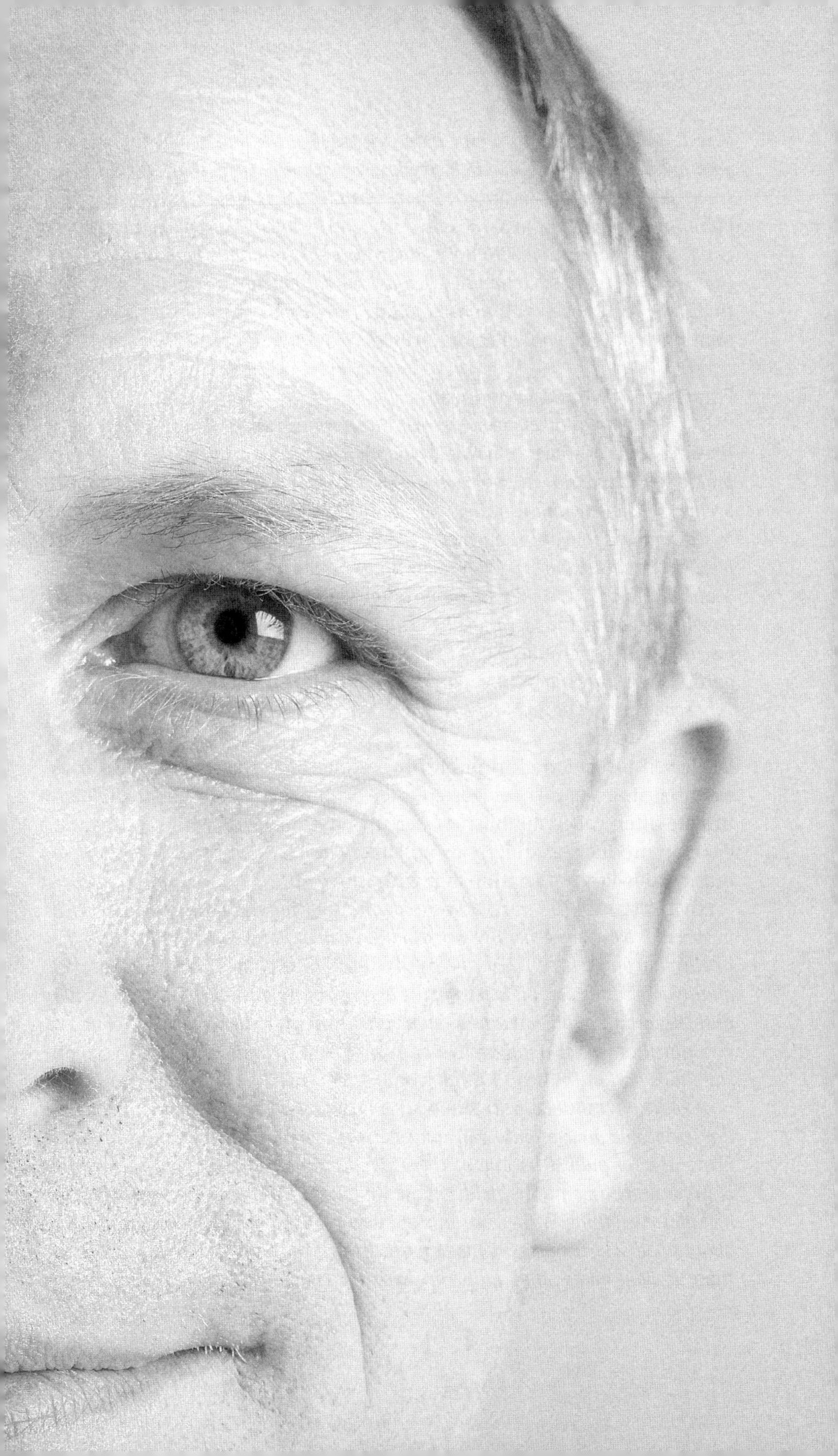

is well understood by our senior management. For me and academics like me who are delivering this research, we know, within reason, that we will have some stability in the medium to long term. That is important in terms of planning and hanging on to the university's really talented people, and it differentiates our cooperation with Rolls-Royce from the standard one-project linkups that we do with other companies. It's really that long-term stability, and only from that, that you get an understanding from commercial business partners that universities are not in it for resources. They're in it for the framework.

Framework? What do you mean?

It's all about that element of real-life demonstration. From a university research point of view, we would probably never build a full-scale, full-power prototype of something we are working on; probably we would build a little one. Seeing is believing and working with Rolls-Royce has given us the opportunity to see our ideas in real-life operating conditions, and that has driven additional research because along the way we've found some interesting issues which we could tackle in a very classic publication manner. You know, if it relates to us without losing the fact that it's also a very good platform for fundamental research, it really gives us a good feeling. It also gives us the feeling that we are part of something interesting, particularly when we get positive feedback from very senior people at Rolls-Royce.

Dr. Jewell just referred to publishing, something that is widely seen as essential in academia and often alien to businesses. Indeed, universities are often called "publish or perish" environments while companies develop products as quietly as possible. Dr. Clarke, how does this play out in Rolls-Royce's relationship with universities?

DAVID CLARKE/ROLLS-ROYCE: We recognize that universities have to publish to succeed. Now, we generally work to a principle that we own and protect the intellectual property (IP) that comes out of the UTCs. That is part of our internal quality procedures, and it is part of the agreement we have with the universities that they won't publish without our at least reviewing the publication. But it's very rare that we ask a university not to publish something, very rare – you can say that's part of the deal. I think it works very well.

GERAINT JEWELL/SHEFFIELD UNIVERSITY: I'd endorse David's comments. Rolls-Royce's people understand what universities want out of the relationship, and being able to publish is one of them. Understanding that is critical. What is important to us at Sheffield is that we operate in a single agreement that includes everything. When we initiate new projects, we don't have to worry about nondisclosure agreements or putting anything particular in place. Everything is done under that umbrella with the consequence that it very rarely causes any problems. It doesn't steer our research, but it is really important to recognize that when working with industry there is a different set of rules in terms of dissemination, in terms of intellectual property, than there is in non-industry-funded research. So, on occasion, we run another little bit of research parallel to make it less specific and generic, knowing already when we begin a project what we will be able to publish.

Investing in networking

You've mentioned the blurring of boundaries, the different cultures, and the benefits of long-term relationships. Long before initiating this project, Rolls-Royce was a leader in aerospace, energy, and marine propulsion. One would imagine that the company already had its own team of internal researchers, methodologies, and its own sophisticated recruiting network for scientific expertise. What then prompted you to actually go out and create a new cooperation model with universities?

DAVID CLARKE/ROLLS-ROYCE: The goal from our point of view was to give us the most effective technical solution in terms of delivery of technology and capability in some very specialized fields. And then of course it was to be affordable at the same time. Now, if you look back to before we started this in the middle of the 1980s, we had relationships with more than 80 university groups, all of which were working around gas turbines. We had a couple of people at Durham and one in Sheffield and another at Cambridge, and so on. Maybe they talked to each other, maybe not. We set out to rationalize that down with a view to create a critical mass of technical skills and provide the facilities to support that team. Then we put them all together in one place, in some cases as many as 25 people working on similar projects, not exclusively, but primarily for Rolls-Royce. Now 25 good researchers – that is a hugely powerful team to bid for funding.

So networking great minds can be seen as an investment in gaining external funding?

Yes, absolutely! If you look at what tends to happen, when any group of skilled researchers comes together, they're obviously going to build more capability among themselves. One of the big advantages from our point of view is when you take the good researchers from a university, they are also working with other companies in the same area, meaning they get to see things from completely different perspectives, which in turn benefits us. That kind of group working in a broad area is a hugely powerful asset to go out and access finance in the broadest sense.

How does the university see that?

GERAINT JEWELL/SHEFFIELD UNIVERSITY: We joined a network of about 20 UTCs within which we collaborate directly. That means they're complementary rather than competitive and that's a nice environment because academia is increasingly competitive. We are now engaged with people I genuinely think we wouldn't have talked to otherwise. And the fact that we're all under this umbrella is a very powerful thing, as David mentioned, for securing funding for collaborative grants. There is also an element of prestige associated with a world-class company. I think it carries weight with funding councils, in government, in attracting graduate students to stay on to do research and even attracts 18-year-olds coming to Sheffield to choose their university; it's a great value to the university.

Is this network, this umbrella that Dr. Jewell speaks of, the key to the UTC model?

*DAVID CLARKE/ROLLS-ROYCE: I think from Rolls-Royce's point of view, there are two things, and one enables the other. It's about people and their relationships. We've been able to get really good working relationships between the leading staff at the UTCs in the universities and our engineers here in the company. Now, it is easy to say, but it's very difficult to pin down exactly what makes that work. I see two things. One is our people. We try to find the right people to run the UTCs from our side, staff who actually understand what goes on in universities. The other factor is that there are very few cases where the senior leaders of the UTCs have left. If you take the first two UTCs that we set up back in 1989–1990, one of the leaders of those two groups retired only last year (2005). So, we've been able to have relationships for over 15 years with the same people. And because the engineering challenge we offer frankly is huge, when the research works, people can say "that airplane is flying with my technology." **I think that the entire package, the right people, the right working relationship, and being able to sustain that for a long time, like ten years or more in some cases, that's what really makes it work.***

But surely the road to the present has not been perfectly calculated. Goals rarely match realization. Can you recount any such stories, both good and bad, that occurred since the beginning that took you by surprise?

It seems astonishing, but after six or seven years, three of our UTCs set up independently but in related fields came to us without any pushing from us, and said: We want to work together as a partnership to support Rolls-Royce. They felt that this would offer a better long-term solution, a more affordable solution to the company, and would give them long-term security as well. And actually, they've called in at least two other groups. They've done that off their own back, even bidding for funding together. That has worked phenomenally well.

One thing that's not gone quite well is where the research interest in an academic team started to diverge from what we needed. We had a particular case where a UTC team felt the particular technical solution to a problem we had was down avenue A, and our guys were equally clear that it had to be down avenue B and had good reasons for it. We just couldn't find an agreement and ended up closing that UTC. It was interesting that the fact that the industry guys were paying wasn't a good enough argument. The differences in opinion became irreconcilable, the academics just said forget it. In other similar cases it has worked out. In one case we actually asked a UTC to change direction 180 degrees, and they did it extremely effectively. Over a space of 12 months, they actually turned their entire program and entire research base into what we needed. That's success from my point of view.

GERAINT JEWELL/SHEFFIELD UNIVERSITY: *You know, we [the Sheffield UTC in Advanced Electrical Machines and Drives] have only been in existence three years. When we started, electrical engineering was still somewhat of a peripheral thing. But in three years interest has taken off. The numbers of programs and demonstrations are growing exponentially and we've benefited tremendously from engagement with Rolls-Royce. I think if we look at where we are three years after we opened, I don't think that we thought we'd be as far along the line of being really embedded throughout Rolls-Royce. By that I mean we traditionally had most of our interaction at the strategic research level. Now it's right out in Rolls-Royce's business units. Now, it sounds strange, but I can't really think of any negative points. We've succeeded. That may be a consequence of being set up at a time of great demand. We've certainly been under no pressure to scale back on anything. Quite the opposite, it's growing, we're as busy as we can be. All of that's been very good for us.*

Certainly, though, many businesses and universities who try to build cooperation models make mistakes or even fail outright. Why do so many of their attempts fail and what would you advise others seeking such cooperation to do?

What really drives research is continuity. A lot of industrial support is structured on a project basis: you know, a three-year program followed by a six-month gap followed by another two-year program. That's not very good for universities in terms of keeping the core of people working in that area. Universities tend to have, by definition, a high throughput of people and the rundown time for a successful research group is around two years. If funding isn't continuous, the wrong people may leave. Short-term projects are fine. They fill a need. We do that within the university too, but that is a very different animal to a long-term research partnership. I think it's far more valuable for the university to have a given amount of money continuously over five years than enormous peaks and then nothing. That sort of boom–bust kind of operation just doesn't fit nicely with research in universities. That's the biggest mistake.

DAVID CLARKE/ROLLS-ROYCE: *I think what we've made every attempt to get good at, and I think that is something many companies don't do, is to make sure that our staff understand the technology, they're not just business managers. We have a very high level of interchange and involvement. Our engineers go in and out of the university regularly, and they're interacting directly with the team in the UTC. Our people are there every week, and in some cases every day. And of course we don't have only one person, there might be half a dozen or more engineers in and out of that research team in a period of weeks or months. I've seen companies that really tried to constrain it to just one industry person actually based in the university. When you do that, the interaction is lost, and you just don't get the necessary depth.*

Big chances for small enterprises

Rolls-Royce is a big company and it seems that its cooperative efforts with universities in terms of time, money, and personnel are far beyond the reach or scale of a small or medium-sized company. It is hard to imagine that SMEs could come up with the resources to establish long-term financial support or spare an engineering team to work regularly with university researchers.

There are a few options that immediately spring to mind. SMEs may actually team up and form some sort of collaboration. If this is not possible due to the competition between them, which is often intense, universities might turn it the other way around and engage SMEs. There are a number of advantages from that point of view. There are a lot more SMEs than big multinationals, so it opens up a much bigger pool of support and offers big potential, particularly if some kind of joint venture can be established whereby the university can gain a stake in the SME, assuming it's a real "S" in the SME. Realistically, a lot of SMEs in the technology area come out of universities anyway. There is also, I think, a big role for government agencies in brokering some of this. Regional development agencies in Scotland and Ireland are attempting to do exactly that, and they are making some progress. But it is difficult; it is not the kind of relationship readily transferred to an SME.

GERAINT JEWELL/SHEFFIELD UNIVERSITY: *From the university's point of view, the issue with SMEs is not the scale of resources. It's all about commitment. It's unrealistic to expect a small SME to put large sums of money into a project. They might be able to partly fund a student or provide assistance in kind. We've worked on very small scales with companies having virtually no R&D, in some cases stimulated by nothing more than someone seeing a paper that we've done. And we've found synergy and a fairly modest way to grow it. Sometimes things just go along at that level because that's all that company can absorb. The scale of it can be fairly modest, but I think the one thing that you can learn from the Rolls-Royce UTC model is that it's about the time that the company is putting in.*

Giving us a contract and then turning up every three months and saying: Where's my stuff? just isn't going to work. It's all about keeping people in the university as part of the team, keeping some involvement, and, actually it sounds strange, but growing a lot of loyalty to that company. And that doesn't come easily. When it is there, that is when researchers will go the extra mile; and that has more to do with a specific mind-set from the industrial partner than resources.

Dr. Clarke suggested that the responsibility to go out and engage might be better placed on the university. Dr. Jewell, you mentioned the possibility of a small company financing a student. Is that something valuable enough for a university to go out and seek? Is that a good place for SMEs to start?

I think so. Universities are under increasing pressure to conduct knowledge transfer, from spinout companies to engaging with small SMEs and multinationals. There are hundreds of cases where it's worked extremely well and where fantastic things have come out of universities; and an equal amount of effort dissipated into things that were never quite set up properly. Now, getting academics enthused to go out and work with SMEs, that's a big thing! There are people like myself who spend a lot of time doing industrial research, but there are also a lot of others who just haven't had that exposure. Trying to show them the value that this has in respect to their research takes some education. Now, there are opportunities, but SMEs need to accept that by the very nature of speculative research, not all projects will ultimately achieve successful outcomes.

On distance, cities, and racetracks

We live in a time of globalization, where components of business are spread over vast distances. Advances in communications technologies play a key role in that. Can such technologies help bridge the distance between academia and companies, particularly if they are located in rural areas?

DAVID CLARKE/ROLLS-ROYCE: Communications technology can certainly help, but can it change it fundamentally? I think, frankly, no! Half the value in the relationship is working with this *university and* this *guy. When you're close by, you can do that anyway. You need that kind of interaction. Could you handle it all remotely? A lot of it, yes; but I think all of it, no! You need to be able to go in and out of the lab regularly, whether it's with one person or ten.* **And there's no doubt that to start a project, you've got to meet the people in person and get to know them, to see how you can work together.**

You both know the phrase "location is everything." Do you really think it matters? Are companies in rural settings at a disadvantage when seeking to cooperate with universities?

GERAINT JEWELL/SHEFFIELD UNIVERSITY: We work with companies in areas that we would regard as being out in the sticks. Before joining the university, I worked briefly for a company that was far into Cornwall, making high-tech magnetizing equipment. It was a three-hour drive from the nearest city. I think that it's more difficult for them than if they were plugged into the motorway network, but that's the choice they made. It's a balance between the quality of life and the support they get locally.

So can one say that cities matter?

DAVID CLARKE/ROLLS-ROYCE: That's a difficult one. I'm not sure that cities matter. It's definitely having a mass of activity going on in one place. Communication matters, that's reality. Here's a good example: Formula One motor racing is generally based in a chunk of the British Midlands between Northampton and Oxford. It's not built round a university. It's built round a racetrack in the middle around Silverstone. That's why it's there, because guys can run tests. But it's also a major technological employer in the region. It brings an awful lot of economic activity into that area, and hence it's got regional development support. As a result, they've built a new road through the middle, which acts as a communications corridor, meaning that Northampton to Oxford has now become kind of a technology corridor for those guys. But it's not built around a city. It's a mass of people in one place with some expertise and skills. Whether it's high-technology, postgraduate skills, or simply the guys who understand the manufacturing aspects of Formula One, it's built around them, around people, and it happened to start up at a racetrack, not a city.

Outsourcing or working together

What would you both describe as the most meaningful achievement or achievements as a result of the cooperation between Rolls-Royce and Sheffield?

I think the greatest achievement for Rolls-Royce has been the fact that we've ended up exceeding our expectations. I measure that by the fact that university technology centers are now an integral part of our research base. They're not duplicated in-house.

Does that mean it's been outsourced?

*No! It's easy to look at it and say it's been outsourced because we don't always do the research ourselves anymore, but I wouldn't call it outsourcing. The reality is it's a "we do this together" type of relationship. Have we brought something in? Yes. We brought in a much more challenging set of research staff than we would have had if we'd done it ourselves. **They challenge us in ways that we couldn't do, if we'd done it all by ourselves. They've come with a different perspective on a whole range of problems, and somehow we couldn't handle it without these guys.***

GERAINT JEWELL/SHEFFIELD UNIVERSITY: *Institutionally, if I look at our UTC in particular, I think we really have the opportunity over the next few years to put ourselves among the leading groups if not become the leader in what we do, and that would not be remotely possible without Rolls-Royce.*

And from a personal point of view?

DAVID CLARKE/ROLLS-ROYCE: *When I go and talk to these guys, I see just the sheer breadth of capability that they can bring, the skill and capability that the university side offers. I recently got the chance to look at what's happening at a particular university and I found myself looking at something that had nothing to do with engines. It's just that chance to see, from a technology point of view, not just what they do but all the peripheral stuff around it. It's just fantastic.*

Thank you for the interview.

Key project insight

- The University Technology Centres (UTCs) are centers of excellence set up by Rolls-Royce at carefully selected research universities. They give the company access to the world's leading engineering talent. For their part, universities get real-life challenges tackling complex engineering problems in stable, well-financed, and long-term environments. Its objective is to create a cross-cultural, strategic relationship with two-way benefits in every sense. Another important aspect is the importance of industrial discipline and academic freedom.

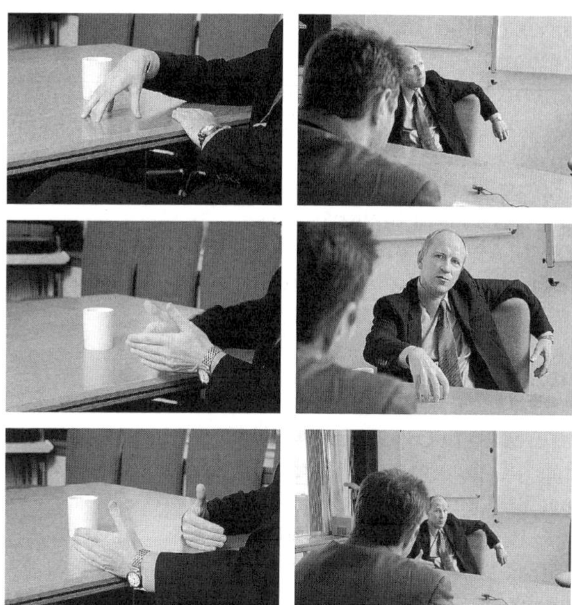

- The key to its success can be summed up in three words: continuity, flexibility, partnership.
- Continuity: long-term commitments, both personal and financial. Continuity builds trust and loyalty, which strengthen cohesion between partners.
- Flexibility: Rolls-Royce and the universities are willing to adapt to the needs and working methods of their partners.
- Partnership: there is regular contact between the universities and Rolls-Royce, which allows ongoing coordination of the common goals.

Basic facts and figures?

- The first UTC was created in Oxford in 1989 with the aim of working on solid mechanics; the second UTC (vibration) was set up a year later at Imperial College in London.
- There are now 26 UTCs worldwide, with 20 in the UK, 3 in Germany, 1 each in Norway, Italy, Sweden, and the USA, as well as an "Advanced Research Centre" in Singapore. Additional research centers are currently being developed in Spain, Korea, Japan, and India.
- The Rolls-Royce UTC in Advanced Electrical Machines and Drives at the University of Sheffield employs 12 researchers.
- Rolls-Royce's UTC network consists of approximately 400 researchers worldwide and operates with an approximate annual budget of over 7.5 million euros. Independent of its UTCs, Rolls-Royce employs approximately 7,500 engineers.
- The number of patents that have come out of the UTC network ranks in the hundreds.
- Rolls-Royce employs 37,000 people worldwide.

Idea Competition Instead of Managing Weaknesses: Cooperation and Regional Development II

Interview by *Martin Rasper*

Sweden

SvD
IZENIT
5

"It isn't so much about competition as it is about helping the participants develop their strengths, focus their ideas."

An original approach for promoting regional development: focus on resources rather than weaknesses. The Swedish Vinnväxt program supports not disadvantaged regions but ones with special strengths. And it specifically furthers the cooperation between universities, business, and administration.

An interview about a prospering cooperation model for economic development in rural areas, based on the example of Triple Steelix in Bergslagen, the traditional mining region in central Sweden.

Triple Steelix focuses on the steel industry as a joint initiative of this region, the steel companies, and the academic sector, and it is promoted by the Vinnväxt program as such. Represented in the project are two universities, the regional administrations of the provinces of Dalarna, Västmanland, and Gävleborg, some ten municipalities, eight large steel companies, and roughly 400 small businesses that offer either services, processes, or products related to steel.

The name Triple Steelix has a double meaning: on the one hand it alludes to the three provinces spanned by the project, and on the other, to the triple helix structure of the Vinnväxt program that is supported by the three pillars of business, university, and administration.

Lars Hansson, born in 1955, comes from Bergslagen, a mining and steel district where, as they used to say, children were "born with steel in their eyes." Hansson studied metallurgy at the Royal Institute of Technology in Stockholm, and went on to work as an engineer and manager at SSAB, Sweden's largest steel corporation, and at Morgardshammar, a steel company with a long tradition of quality. Since 2001 he has worked for the Swedish Steel Producers' Association Jernkontoret, where he has been in charge of running Triple Steelix since 2002.

Lars-Gunnar Larsson, born in 1951, trained as a medical technician and spent many years doing development aid work for state agencies and private organizations in Vietnam, among other places. He later held various positions in development and regional promotion projects in Sweden; Vinnova, the Swedish Agency for Innovation Systems, was founded in 2001, and since that time he has run the Vinnväxt program there.

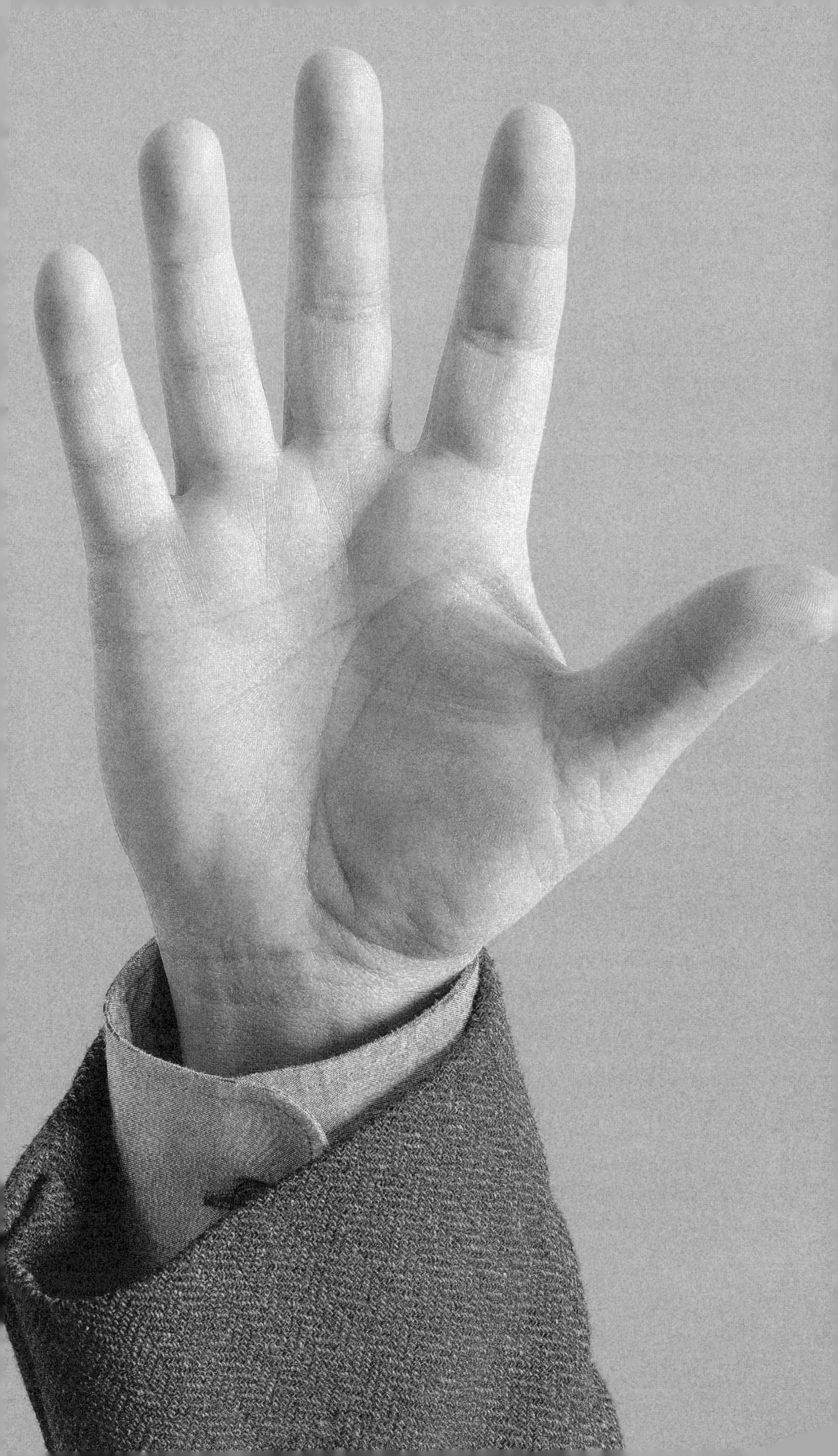

Stockholm in the fall. A cold wind and surreally clear light form the contours of the city and hint at the vastness of the country. The governmental agency for innovation systems, Vinnova, sits in the middle of the bustling city, its headquarters in the same high-rise office building as the daily newspaper *Svenska Dagbladet*.

Lars-Gunnar Larsson runs the Vinnväxt program at Vinnova. He is a serious, reserved man who seems intent on dispersing all notions that his position might entail the least bit of power. Lars Hansson, on the other hand, is on the Board of Directors of Triple Steelix, an initiative of the steel industry in Bergslagen, which is supported by the Vinnväxt program. If Swedish steel needed an image campaign and was looking for someone to run it, Lars Hansson would be the man. He brings together the resolute nature of the pragmatist and the classic European engineer's reliance on theory. With this man at your side, you fear neither a broken axle in the middle of the tundra of northern Sweden nor the dilemma of how to start your dissertation on the plasticity behavior of ultra high-strength steel plate. Hansson speaks with a sonorous, sometimes almost throaty voice that takes on a hearty tongue-in-cheek nuance when he seasons his talk with technical terms in German, my mother tongue: "Ståhlblech," "Dråht," "lågerhaltende Händler," "Verr-ein deutscherr Eisen-hüttenleute." Ah, and he's also a captain in the military reserve – the infantry.

Mr. Larsson, Mr. Hansson, there are two points about the Vinnväxt program that we found especially convincing: first, the fact that all three pillars of industry, university, and public administration – the so-called triple helix – have to be incorporated in each funded project. And second, that each project has to give itself its own infrastructure with someone at the top; in other words, Vinnova wants to have a clearly designated person in charge as its contact person. Is this true?

BOTH: uh . . . yes. No. (laughing)

LARS-GUNNAR LARSSON/VINNOVA: You're right in the sense that we need a contact person for administration purposes, a formal person in charge, but actually . . .

LARS HANSSON/TRIPLE STEELIX: . . . usually all three of us went, actually.

LARS-GUNNAR LARSSON/VINNOVA: That's right. In principle, we don't want just one person at the top. We prefer a team.

LARS HANSSON/TRIPLE STEELIX: Officially, I was the project manager, but the three of us worked together as a team – and a very close team at that.

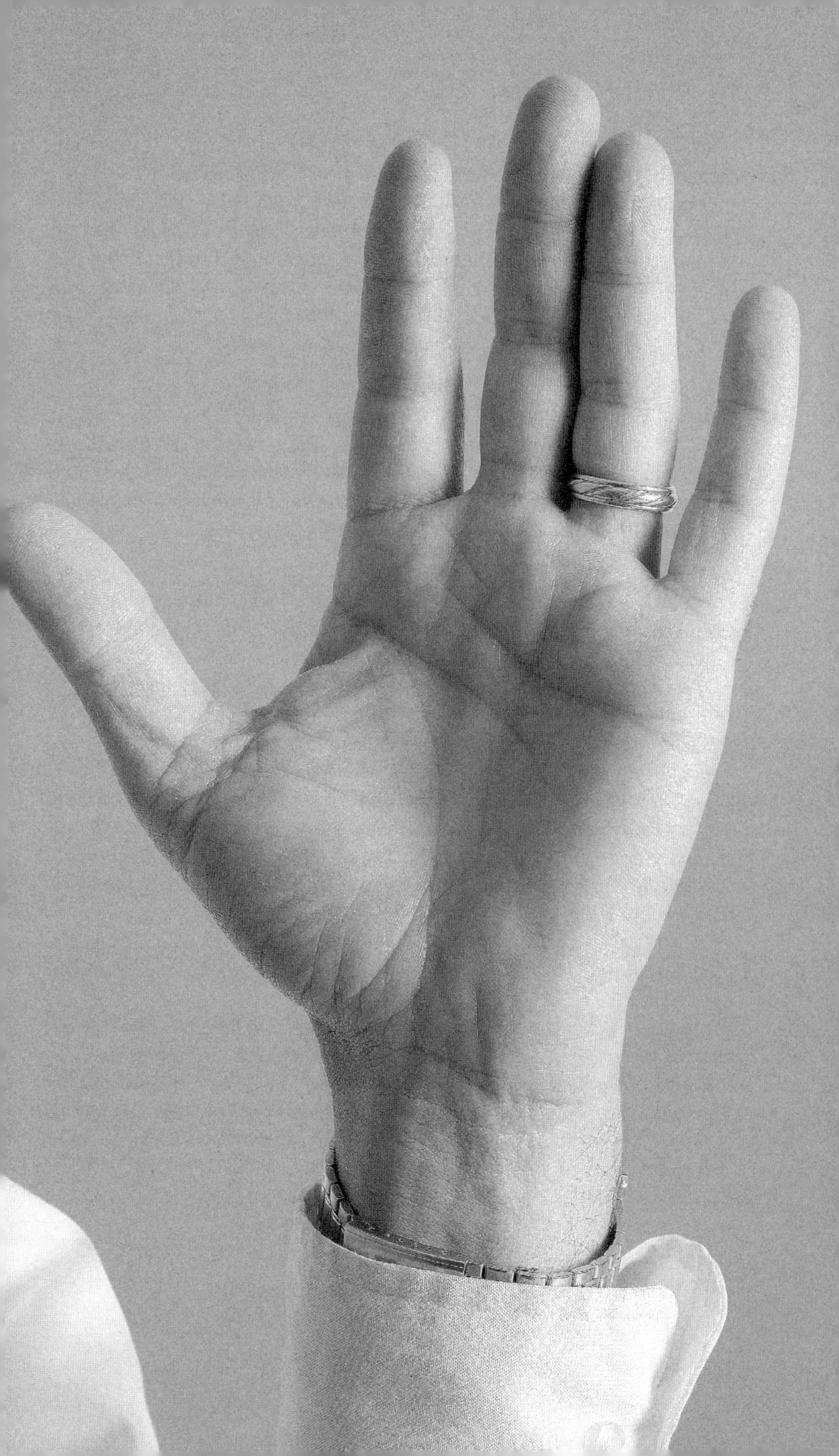

Does this kind of team ideally consist of people representing the different aspects of the projects, i.e. university, industry, administration?

LARS-GUNNAR LARSSON/VINNOVA: Yes, of course, if possible. But sometimes people grow into the project over time and automatically take charge. Or what also might constitute ideal conditions for collaboration is if you had someone like Lars Hansson in the case of Triple Steelix, someone who has worked in both research and industry and then maybe even has a bit of an idea about politics, about how decisions are made in a region like this.

How did you come up with this triple helix structure?

That wasn't our idea, we just borrowed it. The concept of the triple helix comes from the American innovation researcher Professor Henry Etzkowitz and has established itself in the USA as a good structure for promoting innovation processes. This inspired us, and we adapted it for our purposes.

The Vinnväxt program was set up as a competition that was to reward the eight winning regions with concrete funding. What happened to the other ones?

There were several selection rounds with more than 150 participants in the beginning; and at the end there were two final rounds to choose the winners. Triple Steelix, for example, didn't even make it to the first of these two rounds, it had already been eliminated ...

LARS HANSSON/TRIPLE STEELIX: ... which was good for us in the end! Because that forced us to revise our concept. And at that point we also got a little bit of money from Vinnova, 1.5 million kronas (approx. 160,000.00 euros) to continue developing our concept, and on top of that we also received funding from the EU, the provincial governments, and the municipalities, giving us a total of 11 million kronas (approx. 1.2 million euros) to work with. The whole thing was already up and running by the time our application was accepted on the second try.

LARS-GUNNAR LARSSON/VINNOVA: This process went on for almost three years and we supervised it the whole way. It isn't so much about competition as it is about helping the participants develop their strengths, focus their ideas. We conduct seminars with them, teach them the theoretical underpinnings of regional development. Some voluntarily dropped out halfway through because they realized it wasn't the right setting for them.

Photo   (**see page 195**) Lars Hansson, head of Triple Steelix

In the end you were left with eight winning regions. What were some of the main traits and commonalities you noticed among them?

All the regions exhibit what we call an innovation system. This is basically a critical mass of companies, people, institutions where knowledge management on a given subject takes place – you could also call it a network or cluster. **And one fundamental requirement was that it had to entail a strategic idea for modernizing an already existing industry**. *We wanted a combination of these two things. For example, the electronics industry has a long history in Fiber Optics Valley. Ericsson had production plants there for 30 years, which means there was already a strong concentration of expertise, and now they have decided to reinvent themselves with fiber optics.*

Or Robotdalen, which is building its vision of becoming the world leader in robotics on a strong mechanical engineering base that includes Volvo, ABB, Atlas Copco. Or in a different sector altogether you have Uppsala Bio. Here there is an academic tradition in biology with an outstanding academic repu-tation that goes back to Carl von Linné. Two Nobel laureates have come out of here, and now they have chosen to concentrate on biotechnology and cre-ate an industry here. As for Triple Steelix, the key idea was to integrate new services and fields of activities into the already existing steel industry because now traditional steel companies are producing more and more with less and less manpower.

LARS HANSSON/TRIPLE STEELIX: The industry would do fine without Triple Steelix, but it takes part and contributes. Last week, in fact, there was another big meeting with people from the steel industry-related service sector and the managers of all the large steel companies. They meet, make new contacts, talk, develop new ideas. The idea is to strengthen the service sector around the steel industry core. The large corporations still have no choice but to outsource much of their activity – but they at least want to make sure it stays in the region.

Subsidies as an impetus for developing strategies and networks

How specific are the stipulations as to how subsidies are to be used?

LARS-GUNNAR LARSSON/VINNOVA: The money itself isn't what's important, we've always made that clear from the start. The money is basically used to get things rolling, give development projects a little push, etc. The idea isn't to

---

Photo   (**see page 199**) Lars-Gunnar Larsson, in charge of the Vinnväxt program at Vinnova since 2001

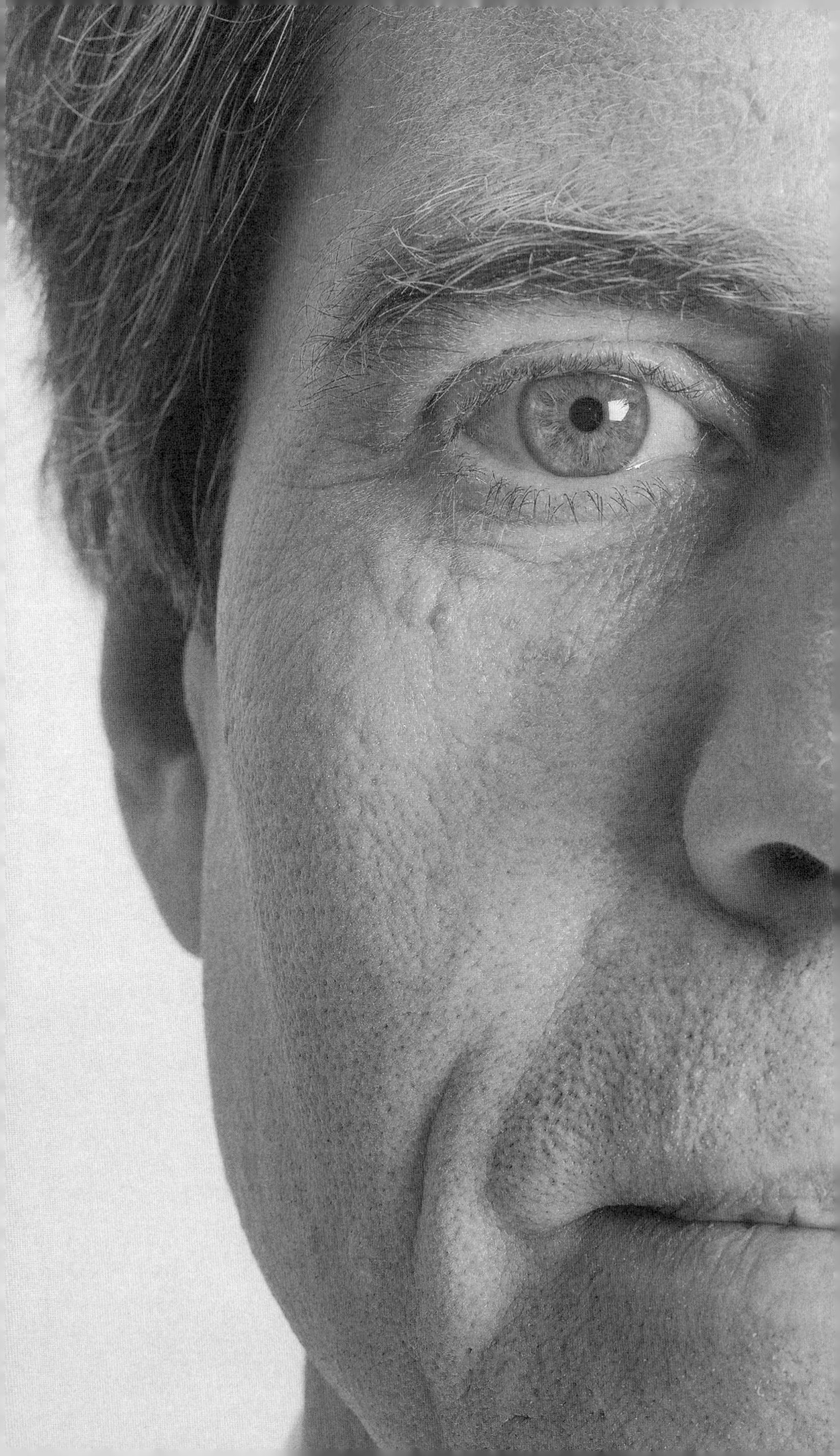

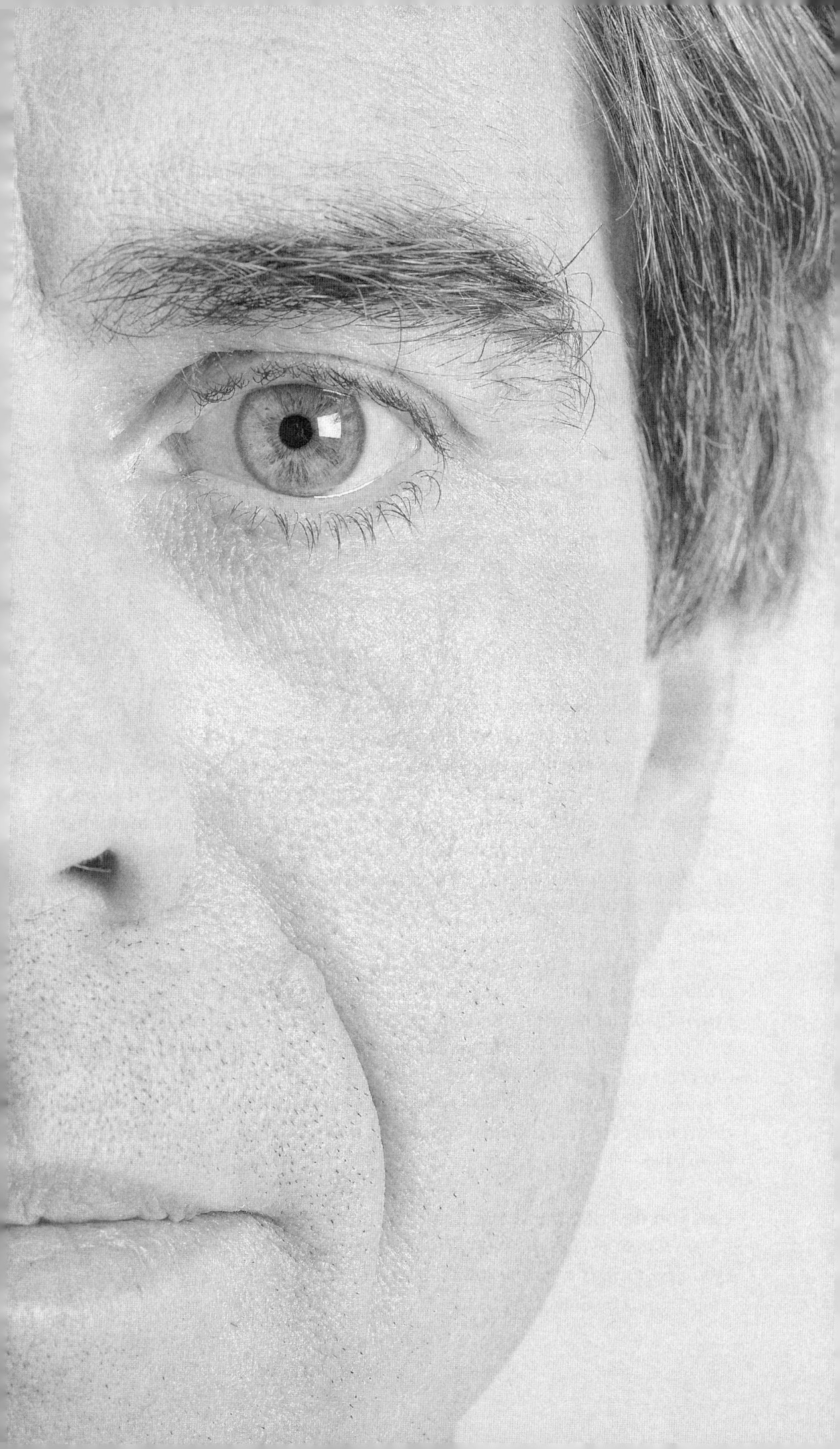

get the companies to mobilize funding sources, that wouldn't work anyway because what we have at our disposal is relatively insignificant. Take Triple Steelix, for example, they get 60 million kronas, that's 6.6 million euros, over a period of ten years. That's nothing compared to the turnover of the major steel companies. Instead, what the money is used for is to set up the structure, the network. How they choose to do that is up to the project managers.

How important was the Vinnväxt program as an impetus to Triple Steelix?

LARS HANSSON/TRIPLE STEELIX: Vinnväxt set everything in motion, without it we wouldn't have founded Triple Steelix. Or put it this way: we might have done something similar, but we would have given it another name. And we probably would have gone about it differently. Just participating in the program exposed us to so many new things: we learned about the administration, the politics, the theoretical background, regional development, cluster formation, and all these things . . . Actually, we were already a cluster, but we just didn't know it. (laughs)

Is the form and size of Triple Steelix typical for the regions in the Vinnväxt program?

LARS-GUNNAR LARSSON/VINNOVA: There is no typical region per se. They are all different. There are different kinds of companies everywhere, each with its own unique relation between research and industry, different geographic relationships. For example, Triple Steelix is unique in the sense that it spans a vast geographic area and because it is centered around a core of large traditional steel companies. Uppsala Bio is completely different. It is concentrated in a city, and the focus is on an entirely different kind of technology that consists mainly in research; the same goes for BioMedicine, where a large part of its activity takes place in Göteborg.

By contrast, with New Tools for Health in Östergötland the focus is on pre-venting illness, home care, the entire health sector including sports – that is an extremely broad range, the network is very heterogeneous and the infrastructure is still loose; it hasn't yet developed the routine that Triple Steelix has, but it has enormous potential for development.

Each case is unique and has to be approached differently. Our clients know this already. We give them the framework and tell them they have to fill it in themselves.

Can you describe the structure of Triple Steelix?

LARS HANSSON/TRIPLE STEELIX: Triple Steelix was set up as a long-term project. Of course there is a formal owner, a formal responsibility, there are legal and administrative matters to deal with. Jernkontoret, the Swedish Steel Producers' Association, is in charge of that. But this isn't the structure supporting the project. The fundamental principle is that everything revolves around steel. Economic growth based on steel. How can we develop the steel industry? How can we develop steel-related ideas and services and integrate them into the traditional industry? That is the idea and it basically came out of the industry itself. In that sense it was good to have Jernkontoret as a kind of mother. I am formally on the Jernkontoret payroll, and in return the Association gets access to the work I do at Triple Steelix as a kind of resource.

Who are the members of Triple Steelix?

In principle everyone in the region who has anything to do with steel. There is no formal membership. Whoever wants to participate is welcome. We have a Managing Board that meets four times a year. It is made up of 13 or 14 representatives from the major areas. We have three provincial governments; at least ten municipalities; eight large steel companies each employing several hundred to several thousand workers, e.g. SSAB, Outokumpu, or Sandvik; roughly 200 smaller production businesses, and another approximately 200 small service-sector businesses – all of them steel-related; and two universities.

What role do the two universities play in the Triple Steelix framework?

The universities in the area are relatively young, both were founded in 1977. The somewhat larger one with a student body of roughly 13,000 is located in Gävle, on the coast, that is to say more or less on the edge of our sphere of activity; the other one is the University of Dalarna, it has approximately 10,000 students and two campuses, one in Falun and one in Borlänge, in the heart of the region. Both universities are involved in Triple Steelix, though Gävle participates to a somewhat lesser extent. Dalarna, for its part, has a representative on the Triple Steelix Managing Board.

It gives it relatively high priority and always sends someone influential, from time to time it was the president himself, at the moment it's the head of the administration department. And usually there is also a professor in each of the different work units. So you see, they want to be there when it comes to making the decisions.

One often hears from other projects that university research and industry have different mentalities, different ways of working, and that they often don't fit together very well. What have your experiences been in this respect?

Yes, sometimes it's a clash of two worlds, for example when it comes to defining the research subject. In industry, research is determined by the needs of the individual companies. The questions are concrete and are closely tied to the real market situation. Academic research usually functions differently. Universities set much more basic goals, draw up a research program, write a grant, and wait for it to be approved. This was sometimes very exasperating, but the university was flexible and willing to try new ways of doing things.

A program for teaching cooperation competency

Do exasperating situations like these happen mainly in connection with the universities or in other areas too?

LARS-GUNNAR LARSSON/VINNOVA: Uff ... We're talking about very complex processes that involve people who may have very little in common, government administrations, dynamic individual personalities, traditionally minded companies, a critical public. Anything is possible. That's why we made sure from the very beginning that we didn't ignore things like communication, motivation, conflict management, and tried, for example, to incorporate them in our seminars for the program participants.

Were these subjects you addressed specifically in the seminars?

Yes, essentially, we held seminars in two basic areas: on the one hand we offered a theoretical background on regional promotion, theory of development models, knowledge management, systemic understanding of processes, as well as on branding and marketing. And then, from branding and setting up your own profile, we went on to the second major area: communication. How do you motivate employees, how do you motivate the public, how do you set up events, how do you conduct media work? During the four or five years of the competition phase we have had a total of 1,200 people attend our seminars.

And as soon as the winners were chosen, that was it?

No, not at all. We continue to work with them on a regular basis, we follow the process, offer help and support where we can. At least once every few months we meet with the people in charge and see how things are going and where there might be a need to make adjustments.

Can we go back to the subject of university collaboration? Did companies and universities work together to draw up degree programs or define the actual subjects to be studied?

*LARS HANSSON/TRIPLE STEELIX: Yes, the university was also willing to make great changes in this area. They both need each other: **university graduates***

need jobs, and the steel industry needs qualified employees. *The university has introduced a master's program in material processing with specialization in metalworking. And it's functioning wonderfully. Last year, I believe 15 out of the 18 who earned their master's degrees found jobs in the steel industry.*

I teach at the university myself. Sometimes I teach three classes: applied material science, metal forming, and production and facility engineering – all as part of the master's program. For lack of time, I've cut back a little, but I still teach my production and facility engineering class.

Have new structures emerged from the collaboration between universities and industry, for example interdisciplinary research institutes?

Structures . . . not so much. Actually, Triple Steelix itself is the structure, and most of what happens takes place in an informal fashion. The idea is to make the city of Borlänge into a kind of center of competency for steel, especially for metal processing and metal forming. Jernkontoret is working to achieve this and so are the university and the companies, and Triple Steelix coordinates most of this process. For example, the Triple Steelix administration office and the Jernkontoret headquarters share the same building with the office of the university chair in materials science. That amounts to close contact. Let me give you a concrete example: the National Graduate School for Advanced Studies of Metal Forming Processes, which I also head, is a kind of seminar for up to ten grad students working on their dissertations on metal-forming processes. Officially, this is part of the university, but it is financed by various sources, by Jernkontoret, the Knowledge Foundation, and of course by the individual companies where the students are conducting the research for their dissertations.

In other words, these graduate students will earn their academic degrees at the university while doing research work at the companies?

Yes, partly at the companies, but partly in university laboratories and workshops too. For example, one of our students did his dissertation for Ortic, a small company that employs roughly 15 workers and produces machines for cold-forming processes. With this equipment a steel band can be given virtually any desired profile – a very specialized, but quickly growing area, important, for example, in construction, in shelving systems for warehouses, in roofs for multipurpose halls, to name just a few applications. For his dissertation the student did research in the interdisciplinary area of the plastic and elastic deformation of a steel band, and this basic research will help the company develop new products.

An ongoing workshop for the business, research, and political sectors

Do the companies themselves approach university institutes, do they have special requests or suggestions for certain topics?

It goes both ways. We have what we call local competency nodes, i.e. informal centers concentrated around the large steel companies, not just in Borlänge. They are kind of like committees. Large companies usually have an extra contact person for this purpose. Within these groups, people have access to the company, to experts for very special topics, they can submit requests, etc. That's the one aspect. And then you have concrete collaboration on development projects: we look for resources we can tap, find out who can do this, who knows that, what data or research results can be found where, and locate where there is the potential for synergy. The idea is to generate as much interaction as possible beyond the classic customer/contractor relationship – as much as possible free of charge. Basically what's going on here is classic networking, people making contact across all fields and disciplines. And this produces new ideas.

That means that these groups are a small-scale reflection of the structure of the overall project?

You could say that, yes. Take the steel-plating sector, for example. The companies here form a kind of cluster and you have a committee with initially one Triple Steelix representative holding it all together; then someone from the steel company is assigned to this task and that person tries to build contacts to the smaller companies, exchange information, develop ideas, give advice – usually that person is someone with a lot of experience in practice, application, and customer contact. And then you've got the research people from the universities or other research institutes; then the government administration folks, that might be the mayor or someone from the provincial government; of course in the best-case scenario it would be someone in charge of economy or industry so that you also had direct ties to politics. And everyone is working together, but the only person who is being paid directly is the Triple Steelix representative.

That means this infrastructure is basically set up and maintained by Triple Steelix, and that's where funds from the Vinnväxt program are being channeled?

Yes. That's what Lars-Gunnar was referring to earlier when he spoke of creating a framework where everything can take place. Part of the money from Vinnväxt also goes directly into research and development projects. The executive committee defines certain priorities – sort of like setting the strategic course for development. For example, we're working intensively on new processes in the area of cold forming because this is very important in the global market.

What role do the smaller businesses play here?

Small businesses play an important role because some of them are very innovative. We have a lot of companies working on new metal-forming methods, for example hydroforming – that means they develop new processes and then create new products which they manufacture based on these processes. Or we have a company that came up with a new way of building mobile phone masts that are extremely light and stable at the same time, and it developed an entirely new production method. So all this is also about creating an atmosphere in which new ideas, new companies can emerge, and then once they've been launched, helping them survive. For example, by letting them profit from the opportunities usually only open to large companies.

How does this work in practice? Is it all very informal? Do the large steel companies let the small businesses use their laboratories and workshops for free? Or do they send the bill to Triple Steelix?

Well, it's like this ... (with a big wave of his arms he gets ready to launch into a long-winded answer, but then he sighs and slumps in his seat) ... It depends. (laughs)

Often there is no charge for small, informal requests. We don't want a lot of red tape. If we're dealing with a larger project that requires decisions by the local committee or an executive committee, the partners generally draw up an agreement, which they sign before they get started. But if a small company needs something that can be done quickly, then it's simply taken care of, and that's that.

Processes rather than structures

And this works because everything operates under the Triple Steelix idea?

*Yes, it works because of the overall idea, but more than anything it works because of the people. The people have to give life to the idea. **The large steel companies only send people into the process who understand and are also interested in the idea.** And it isn't as if we were completely restructuring the steel industry. The structures are there and we get access, we get support. And the reason why, is because in the end everyone profits from the system, even the large companies.*

LARS-GUNNAR LARSSON/VINNOVA: I'd like to emphasize a point we mentioned earlier: we've been talking a lot about structures here, but we at Vinnova prefer, actually, to speak of processes. You mustn't forget that these people, like Lars, have very little power. They have a little bit of money to distribute, but that's all. Their job is to get people to act, to inspire and impel them – but they don't do it as managers, who can give orders, who have power; they are more like motivators.

Are there things where you would say: This didn't work optimally, if we had it to do again, we'd do it differently today?

Not that many. Generally speaking, we are very satisfied with how everything has turned out. One thing, perhaps, if I had to think of something: instead of just defining and focusing on the project itself, we would place more emphasis on integrating it into the social environment from the start. Systematically, on every level: What role should the project play in the region? How do we communicate this? How do we get people involved? How do we get them to see themselves as part of the project?

Would you also regard this as an essential part of the program?

Yes … the process-like character of it. The systemic approach. The under-standing that we're dealing with a dynamic process in which everybody's contribution is important to the success of the project – and in which the power relations and way the participants see themselves are constantly subject to change. It is a dynamic process facilitated by the process-control directors.

In your role as a facilitator do you sometimes have to settle disputes?

LARS HANSSON/TRIPLE STEELIX: Real disputes only rarely, but in principle, yes. Constantly. It's not always simple. It may be easy to understand what we do, but that doesn't mean one can put the concept into practice. As to the researchers and the different mentalities mentioned earlier, we have a special team to explain the Triple Steelix concept and our way of working to the researchers. But we can't force anyone to do anything. We don't have any formal power.

You have a vision.

We have an idea. And our task is to get other people to share this idea and take the same path.

But that's something that can't be learned. Either you've got it or you don't.

We've got it. (grins)

Thank you for the interview.

Vinnova

launched the Vinnväxt program in 2001 as one of its first funding programs. Its goal is to identify regions that stand out for their specific dynamics and an above-average potential for growth. As part of the Vinnväxt Program, eight regions were able to implement their concepts. All regions can access the so-called innovation system, a mix of existing know-how and a clear development idea. A brief description of the eight individual regions:

ProcessIT Innovations: Luleå / Umeå, northern Sweden

ProcessIT Innovations focuses on custom-designed process engineering and IT support for the mining, pulp and paper, and metal-processing industry. In addition to the above-mentioned, the network also includes the universities in Umeå and Luleå and IT product partner companies in Västerbotten and Norrbotten, among others. Its activity produces significant growth due to large-scale, coordinated performance by all interested parties including the commercial players in the region.

GöteborgBIO: western Sweden

Western Sweden and the Göteborg region feature two already well-positioned biomedical clusters. The two fields of focus are biomaterials/ cell therapy and cardiovascular and metabolic science, including diabetes, obesity, and strokes. The aim is to create a good foundation for long-term growth in biomedicine. This will have an impact on both health care and the job market. The basic idea is that academic cutting-edge research leads to innovations in the industry, which in turn give rise to practical application in the public health system.

Triple Steelix: Bergslagen

The steel industry in Bergslagen is the world leader in its niche. Based on the steel industry, a cluster of companies dealing in steel production, metalworking, processing, and knowledge-based services has developed. It works with universities to achieve its goals, which include the development of expertise in the fields of materials, steel processing, nanotechnology, industrial IT, and environmental and energy effectivization, among others. Triple Steelix is backed by such large companies as Sandvik, Outokumpu, and SSAB.

Fiber Optic Valley: Hudiksvall

Fiber Optic Valley is the country's greatest investment in building expertise and production for the application of optical fibers. The long-term vision is to make optical fibers the standard material for all IT communications. Mitthögskolan University, Acreo Fiberlab, and the Ericsson Network conduct joint research in Hudiksvall in the field of fiber optics. Hudiksvall is also home to a test-bed laboratory where healthcare professionals use fiber optical networks for conducting medical communication with patients.

Hälsans Nya Verktyg (New Tools for Health): Linköping / Norrköping

Hälsans Nya Verktyg is an initiative that focuses on regional development in Östergötland by creating individually adapted health solutions. It develops new solutions and growth potential in the areas of preventative medicine, personal care, personal health, and sports. Among its partners are roughly 60 companies including Saab, the municipalities, parliament, and administration of the province, the regional development council Östsam, Linköping University, and other research companies and groups of interested people.

Uppsala Bio

Internationally, the biotechnology sector in Uppsala is well positioned in the field of biotechnological research. It has produced successful innovations in the areas of pharmaceuticals, diagnostics, and medical devices. Internationally recognized trademarks such as Pharmacia and two Nobel laureates positively influence the region's potential for being able to globally market its innovations.

Innovationer i Gränsland (Innovations in Border Areas): Scania

Located in the Scania Province, Innovationer i Gränsland is an innovation system for food. The strategic idea is to raise the return on investment in agribusiness and to create the "health food of the future" with high added value. Creativity and innovation come from a background of pioneering interdisciplinary research. One of its main aims is to develop know-how in its priority field of "nutrition and health."

Another area of focus is healthy and nutritional foods in schools and hospitals.

Robotdalen: Robot Valley / Lake Mälaren Valley

Robot Valley's vision is to become the leader in the research, development, and manufacturing of industrial robots, field robotics, and robotics for medical and health care. The key to success has been an environment where factors such as strong research, qualified education, and industry generate synergies, and where innovation and company start-ups are encouraged. They have succeeded in mobilizing groups of interested people from the entire region. Large companies such as Atlas Copco, Volvo, and ABB are just some of the large companies backing the investment. It is also the site of Sweden's first full university course of studies in robotics, which was launched in the fall of 2006.

Key project insight

The principles

- It's not about subsidizing, but about helping people help themselves.
- We don't promote weak, underprivileged regions, we promote ones with good chances for development – and even then, only if they have a clear concept.
- The strengths and development dynamics of a region are supported by the three pillars of industry, politics, and academics (triple helix), which only work if these three sectors have been consciously networked and contacts established.

Implementation

- Strong emphasis is placed on the process-like character of the program.
- The main focus is on working out the specific strengths and unique qualities of each region.
- Support and further training for the participants through a wide range of subjects offered in courses and seminars on regional promotion, knowledge management, branding, marketing, communication, motivation, etc. with a total of 1,200 people attending seminars during the course of five years.

- Organization of the program as a competition, but without placing too much emphasis on the competitive character. This means even those who don't win continue to receive guidance.
- Conscious integration of surrounding society, whether or not it will produce concrete action.

Using Lost Knowledge to Found Cooperatives:
Cooperation and Development Aid

Interview by *Renata Schmidtkunz*

Turkey

"And instead of moving to the slums of Istanbul and taking poorly paid jobs, now people can stay in their villages and still make a living. They determine their livelihoods for themselves."

A German chemist set out to learn the centuries-old natural dyeing methods used by carpet weavers in Turkey. In cooperation with Marmara University in Istanbul he and his wife researched this forgotten knowledge and started a one-of-a-kind economic aid project.

In some 30 villages they developed a participatory, cultural, and economic model that now lets hundreds of families make a living. DOBAG is an academic–artisanal cooperation project in western Anatolia.

The DOBAG Project (Natural Dye Research and Development Project) is a collaboration between Marmara University and two carpet cooperatives in western Anatolia. The preliminary work for the project began back in 1981. Its official inception, marked by the signing of a formal contract, took place in 1983, thanks to the initiative of the German chemist and biologist Dr. Harald Böhmer.

Dr. Harald Böhmer taught from 1960 to 1967 and from 1974 to 1979 at the German high school in Istanbul. It was during this time that he started his research on natural dyes. Together with his wife Renate, he traveled through Turkey to chemically analyze the color intensity of vegetal dyes that had been used on carpet wool for centuries and to later test them in a series of experiments.

Harald Böhmer concentrated on two regions with long carpet-weaving traditions: the village of Ayvacik and the Yuntdag region, both in western Anatolia.

He spent years researching natural pigments, traditional dyeing methods, and knotting patterns that had been used for centuries. He then passed this knowledge on to the women in the carpet-weaving villages. In order to ensure the production and marketing of the valuable carpets and to guarantee the artisans a steady income and raise their living standards, Böhmer founded two cooperatives and enlisted the cooperation of Marmara University.

From 1981 to 2005 Harald Böhmer worked for Marmara University in Istanbul, and with the help of his colleague Dr. Serife Atlihan and a team of assistants, he supervised the DOBAG Project. Now Dr. Atlihan, who has meanwhile become a full professor, runs the entire project.

Today the cooperatives are made up of some 400 people in almost 30 villages. The Yuntdag cooperative consists of female members

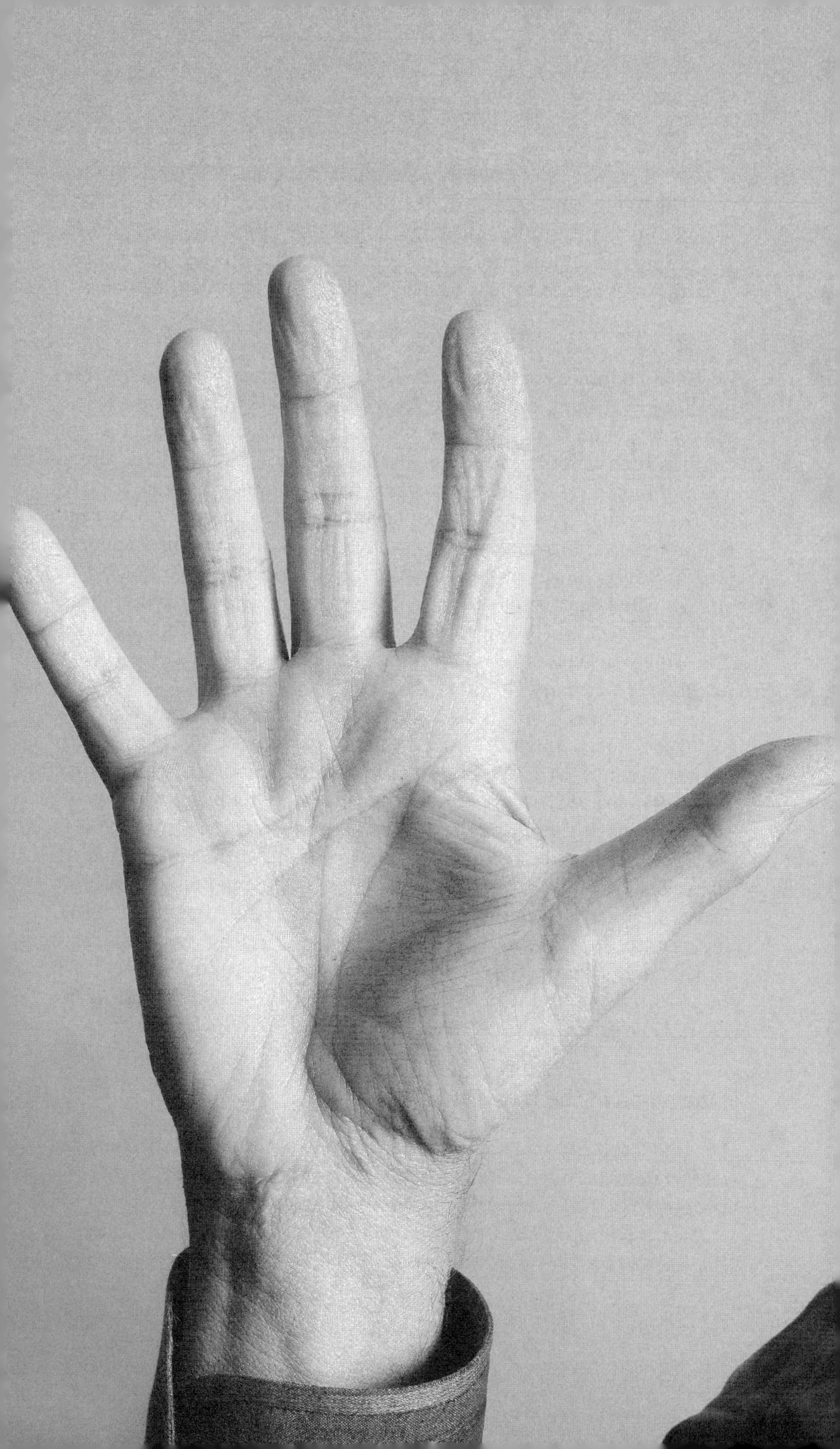

only. Per year some 2,200 carpets are sold to selected dealers in the USA, Japan, Germany, Norway, Denmark, Australia, and Great Britain.

The living standards of the carpet weavers and their families have gone up significantly, and migration to the cities has been curbed. The latter effect was one of the main reasons that the regional Turkish authorities agreed to support the DOBAG Project in the first place.

Dr. Harald Böhmer was born in 1931. He studied chemistry, biology, and physics in Freiburg, and earned his PhD in biochemistry. After teaching for ten years at various high schools in Germany, he taught at the German high school in Istanbul. During this time he started his chemical and biological research on natural dyes. In 1980 he began working for the German Development Service and received a position as visiting professor at Marmara College, which was to become the Faculty of Fine Arts at the newly founded Marmara University in 1983. In 1981 he and his wife Renate and colleagues from Marmara University started the DOBAG Project. In 2005 he retired from his position as visiting professor. The same year he was awarded the Federal Cross of Merit of the Federal Republic of Germany.

Renate Böhmer, born in Wałbrzych (Silesia) in 1931. In August 1946, expulsion/forced migration from Silesia and arrival in Delmenhorst (Lower Saxony). In 1952, graduates from high school in Delmenhorst. From 1952 to 1955, earns her teaching credentials for primary and grade school levels at the teacher training college in Bremen. From 1955 to 1958, school teacher in Bremerhaven. In 1956, marries. In 1957, birth of her first child. From 1958 to 1960, school teacher in Bremen. In 1960, leaves her teaching position behind and moves to Istanbul. In 1961, birth of her second child in Istanbul. In 1962, birth of her third child in Istanbul. Lives in Ganderkesee (Lower Saxony) from 1967 to 1975. In 1975, moves back to Istanbul. From 1981 on, collaboration in the DOBAG Project.

Innovation of the forgotten

Dr. and Mrs. Böhmer, your initiative is responsible for the ongoing cooperation between Marmara University in Istanbul and two carpet cooperatives in western Anatolia. How did this come about?

HARALD BÖHMER: Back then, Marmara University was still the College of Applied Arts based on a German model (Ulm). In the beginning, the teaching

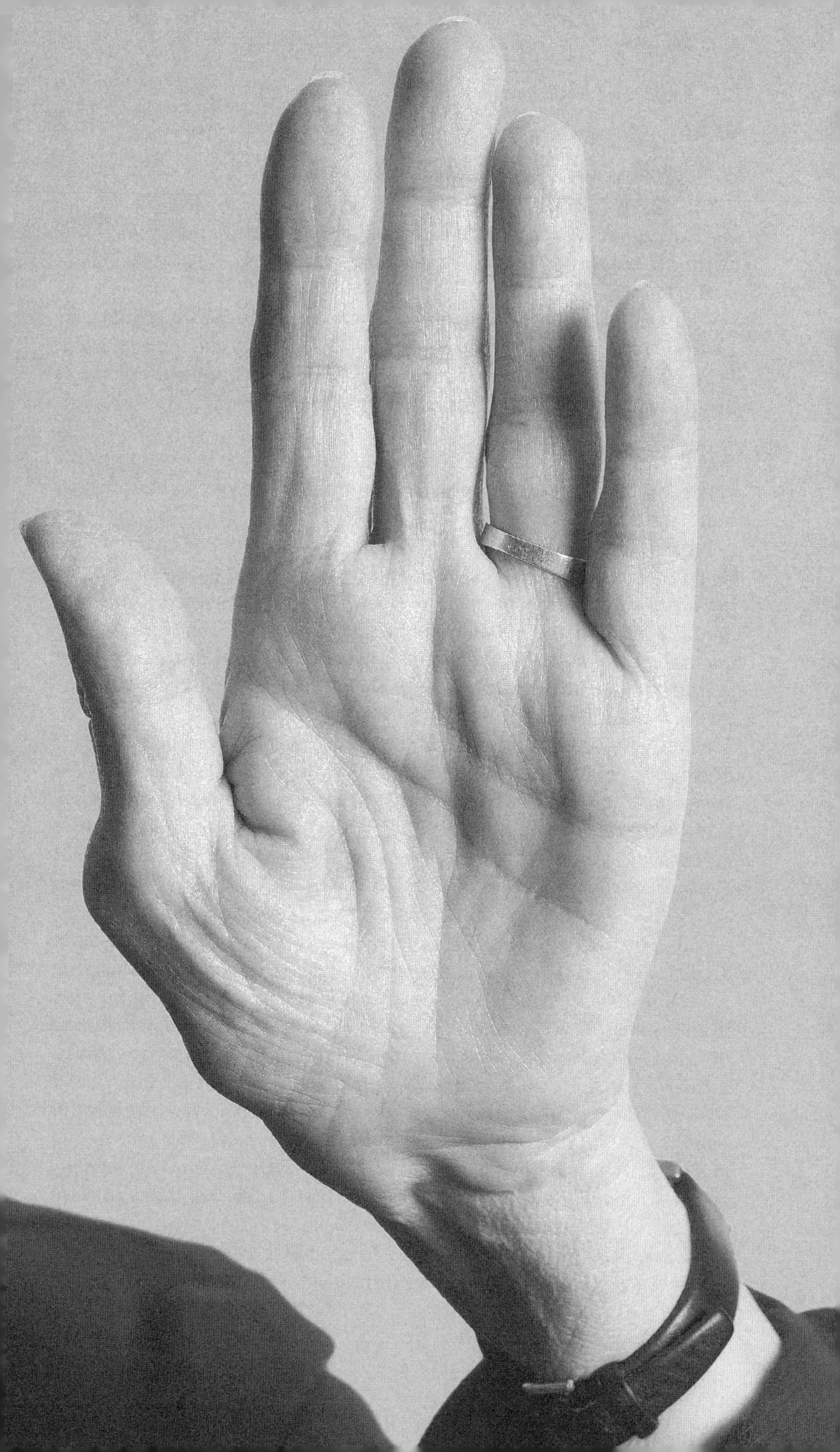

staff came from Germany. When we got there in 1981, there were still several German instructors. Today the whole faculty is Turkish.

This college was interested in preserving traditional Turkish handicrafts and even before we got to Istanbul it had already made initial contact with existing cooperatives. But these organizations made demands that the college couldn't meet.

RENATE BÖHMER: *From the beginning it was an initiative on a private level based on a college instructor's interest in a traditional craft still being practiced in the villages. It didn't matter if it was embroidery, or knitting, or crocheting. It was the craft that the college was interested in, not necessarily carpet weaving.*

Was the university interested in cultivating cooperation between the university and the artisans' cooperatives in order to preserve old traditional handicrafts?

Yes, but they were also aware from the beginning that the quality of the synthetic dyes being used throughout Turkey was poor and the colors didn't go together. And yet even in the most remote villages people preferred synthetic to natural dyes. Since the mid-nineteenth century, knowledge about how to make and use natural plant-based pigments was gradually being lost. And this, of course, had a negative impact on the quality of artisanal products of all kinds, whether knitted, crocheted, or woven.

HARALD BÖHMER: *In the seventies I had already begun reviving the old art of natural dyeing for my own interest. My wife and I traveled from village to village, talked to the old folks, and analyzed the fibers in ancient carpets. Then I set out in search of the plants used for dyeing back then, and through many experiments I discovered how to prepare the dyes and stain the wool so that I could recapture the old color range. Then we went back to the villages and taught these old traditions to the women. I knew I would need funding to be able to carry out the project. The first place I turned to was the GTZ (German Technical Cooperation), a federal cooperation and development aid enterprise based in Eschborn, in the Federal Republic of Germany. They told me my plan to set up an artisanal production operation in the villages around Ayvacik and the Yuntdag region using old patterns and natural dyes to weave carpets was a crazy idea, and they wanted to have nothing to do with it. After that, I contacted the Federal Ministry of Economic Cooperation and Development in Bonn. I was lucky because one of the officials there had a soft spot for old and new carpets and was willing to listen. In the end he promised me funding, provided I found a non-private-sector partner in Turkey. We talked to every organization we could think of and flew back and forth several times. The German embassy in Ankara had to give its approval every time we approached a potential partner. One of the first institutions we went to was KÖY COOP.*

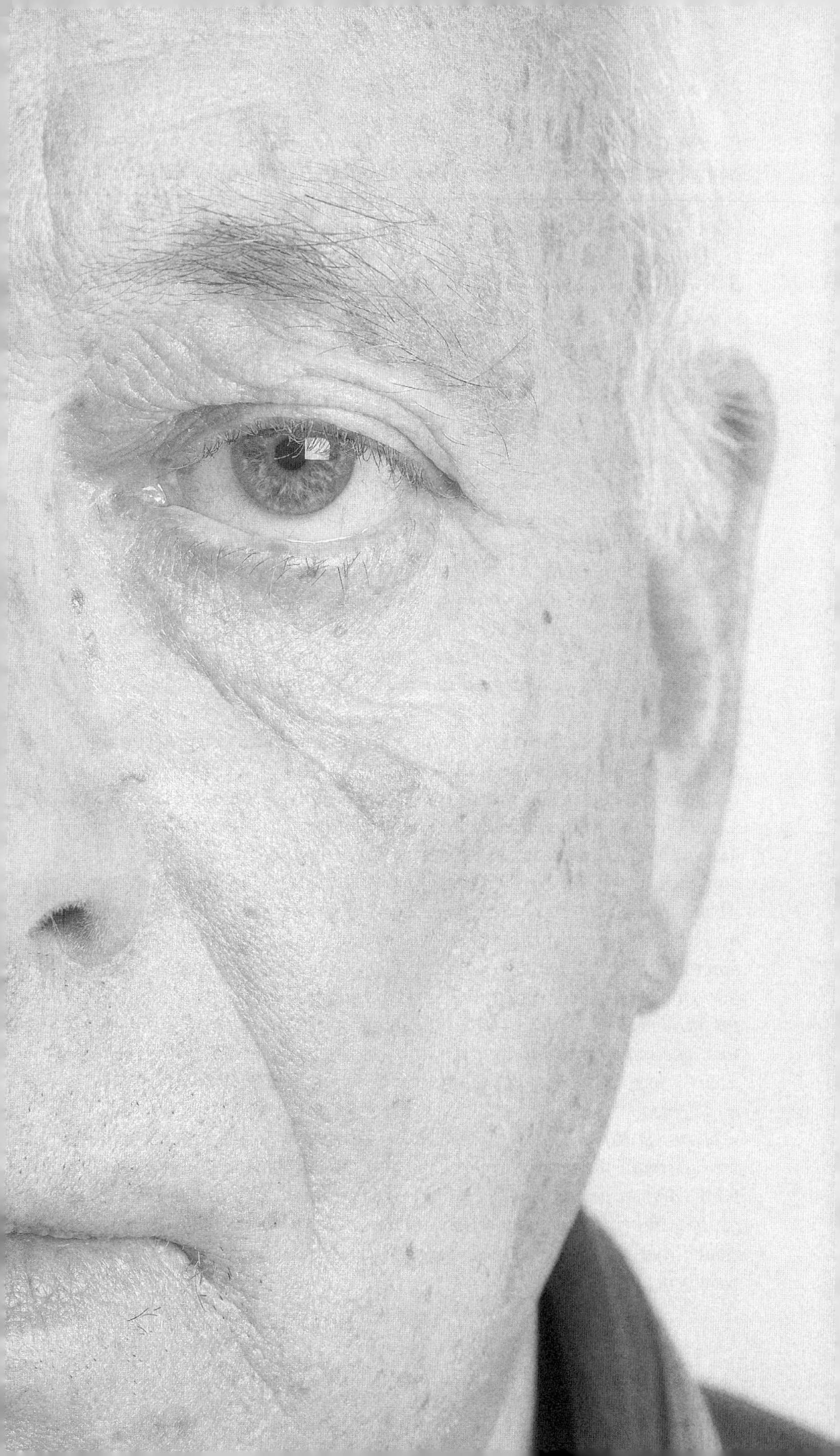

But the German embassy told us: "This Turkish cooperative has such strong contacts to the Soviet Union." (We were still in the middle of the Cold War) "It's crawling with communists, we can't condone it."

Then we went to the Turkish Development Foundation. They were all well educated, everyone spoke English and had studied abroad, but when we went to the German embassy, they said: "Didn't you notice, they're all Americans and Israelis, all secret service people. We're not getting mixed up in that." (laughs)

And then a personal contact put us in touch with Marmara College in Besiktas [district in Istanbul], which later became one of the faculties of Marmara University.

The College of Applied Arts in Marmara – as it was called back then – had always been interested in preserving old crafts. As far as the German development aid authorities were concerned, my Turkish partner could be anyone as long as he wasn't a private carpet dealer because they feared – and rightly so – that this might affect both the quality of the carpets and the weaver's working conditions. Besides, the development aid funds were not intended to be used to finance private businesspeople.

In 1981 I was given a teaching position at Marmara College since that was another condition set down by the German side: **I was to be incorporated in the Turkish system and paid like a Turkish teacher.**

What exactly does the cooperation between Marmara University and the two carpet cooperatives entail?

The aim of these cooperatives is to create a competitive product on a global scale. Incidentally, that was another condition set down by the German authorities, the Ministry. What we produced had to be something that not only rich Turks could afford, but that could survive in the global market. At the time Turkey was flat broke. And the project, which was also funded by Turkish government agencies who put up a loan of today's equivalent of approx. 40,000.00 euros to start the first cooperative, was supposed to help bring foreign currency into the country and curb the enormous migration from the rural regions, where we were working, to the cities. And it did. Therefore, one important aim from the start was to get the cooperatives their own licenses so that they could export directly and no money would be wasted on middlemen.

RENATE BÖHMER: *That was important because if we had left that to a private businessman, the project would have failed. He would have pocketed all the profits and the people would never have seen any money. One of the stipulations of the Ministry of Economic Cooperation and Development in Bonn was to ensure that the money earned flowed back to the villages where the carpets were being produced.*

Incidentally the Ayvacik cooperative completely paid back its loan in just six years. And instead of moving to the slums of Istanbul and taking poorly paid jobs, now people can stay in their villages and still make a living. They determine their livelihoods for themselves. Above all, the situation of the women has improved significantly. That was a big focus for us. In some villages the men have come to accept the fact that now not they but their wives are the ones supporting their families.

Research, teaching, financing . . .
and a spin-off that operates as a self-run cooperative

But what is Marmara University's role in this cooperation project? Do the university and the cooperatives have a contract with each other?

HARALD BÖHMER: We signed the first contract in 1983. That was necessary in order to ensure ongoing work with a long-term development perspective. The university – incidentally, the project is part of the Institute of Traditional Turkish Handicrafts – assumed certain responsibilities. First, continued research in natural dyes and dyeing processes. Second, and this requires the most energy and staff, visiting the villages regularly to check on the weavers and dyers. My colleague Dr. Serife Atlihan is especially active here. She comes from a nomadic family and has been exposed to rug weaving since she was a child. The women trust her opinion and respect her judgment, which in the regular quality-control inspections, without which there would be no quality certificate, can sometimes go against the women, or rather their work. Each rug is inspected for knot count, dimensions, form, pattern, and color quality. The university's third task is indirect marketing; in other words it contacts international carpet dealers who have also won our trust over the years. We only want to work with people who appreciate this natural approach to making carpets and who also support the idea of using these means to help village people improve their standard of living. The cooperatives, for their part, commit themselves in this contract to producing a given number of carpets and paying 4 percent of their sales to the university to cover ongoing research.

RENATE BÖHMER: In the meantime, it is no longer necessary to instruct the women in the villages. They adjusted to the old artisanal traditions very quickly and now weave the old patterns by memory. Some – at least in the Yuntdag region – even make their own dyes. The Ayvacik cooperative employs male dyers and staff to keep their books. The difference between the two cooperatives is that the women in the Ayvacik cooperative own their own materials and basically all the cooperative does is "just" sell the carpets. With the Yuntdag-region cooperative, which only allows female members, all production equipment and materials are jointly owned, and the members meet regularly and make group decisions about new investments and how much of the profits to distribute to the members at the end of the year.

Even though both cooperatives have become very independent, it is important to them and us that an expert from the university drops by once a month to make sure everything is okay. *A lot of times the women just want a little advice about which patterns are selling well at the moment or which colors are popular among buyers. The university staff are their link to the big outside world.*

What exactly do the quality-control inspections entail, and who thought up the rules for cooperation? After all, you are German and not western Anatolian, and at first you might not have been familiar enough with the cultural customs. Under the circumstances, it's easy to get off to a wrong start.

HARALD BÖHMER: We had been living in Turkey for a while by then and could speak the language. That was very important because without language, you're lost. My wife and I had also studied Turkish history and done a lot of research on the history of the carpet.

During our research trips to the villages, we had already made a lot of contacts, and these people welcomed us with open arms – almost gratefully. The rules governing the DOBAG Project arose virtually on their own out of necessity and, of course, in cooperation with my Turkish colleagues. In this context let me again mention Serife Atlihan. She has been part of the project since 1983 and plays a very important role, in part due to her own cultural background.

As to the quality-control inspections, first we determine knot count. ***Over the years we've gone from paying by square meter to paying by knot count.*** *The more knots you have, the higher the quality. Then we check to see if the workmanship of the rugs is technically up to standard. We perform the color analysis part in the lab. If the rugs pass these inspections, they can be put up for sale and receive a guarantee stamped on leather acknowledging that the rug was produced using natural dyes and as part of the DOBAG Project. This is a trademark recognized internationally, and that's a good thing because there are sneaky dealers who try to use our name, a mark of quality, for their usually inferior carpets. But so far we have always managed to deter them.*

RENATE BÖHMER: Each carpet is also given its own number and registered in our archive. Today more than 30,000 carpets have been registered and sold. In addition, we also photograph each carpet. This is especially important to make sure nothing gets lost. And another aspect is that a lot of business with dealers is now conducted via the Internet, so images of the rugs are needed for online viewing. The market took a nosedive following 9/11 and the attacks on the World Trade Center. Some dealers were afraid to come to Turkey. The same thing happened when the war in Iraq broke out and when al-Qaeda launched the 2003 Istanbul bomb attacks that killed some 60 people. But through phone calls or in person we were able to convince the dealers there was no cause for alarm. And now even carpet sales are conducted via the Internet.

HARALD BÖHMER: Our photo archive has grown quite large. Each carpet has a number – as mentioned – and next to it we make a note of the weaver's name and whom/where the carpet was sold to. In addition, the university also organizes international symposia intended to publicize the DOBAG Project in the academic world. There is now keen interest in researching natural dyes. Recently, I was in Hyderabad at an international conference organized by UNESCO. It was attended by 800 experts from 60 countries who came to exchange data and research findings on natural dyes. So you see, there is no lack of interest. Marmara University also initiates articles and maintains international contact with art historians and textile researchers. And naturally the DOBAG Project profits from all this too.

Publish, celebrate, and sell

After more than 20 years, what would you say are the strengths of the cooperation between Marmara University in Istanbul and the two DOBAG Project cooperatives?

HARALD BÖHMER: One of its biggest strengths is that DOBAG is a grassroots model, that means we are directly connected with the producers, and there is no big organization like a chamber of handicrafts between us. In this context it is extremely important to maintain a good rapport between the university staff and the people in the cooperatives. In the beginning we sometimes had problems because our staff, urban Turks, wanted to do the thinking for the villagers, and in this way they disempowered them. But in time that problem took care of itself. Meanwhile, partner relationships have developed, some almost like friendships. Through direct contact it is usually possible to make any necessary course adjustments right from the start.

And since the villagers trust us – because over the years they have seen how our support helps them improve their living standards – they listen and respond to our constructive criticism accordingly. They don't drag their feet.

What is done to cultivate these personal relationships? Do you and your co-workers just show up in the villages for inspections or are there also occasions for celebration and personal exchange? You yourself mentioned the cultural gradient between urban and rural Turks.

RENATE BÖHMER: We encourage opportunities like these and even organize special events. We celebrate with the members of both cooperatives once a year – once in Ayvacik and once in the Yuntdag region. We meet in the nearby woods;

all the women from the villages come, and we sit around and talk. The men prepare and serve the food for these occasions. There is an awards ceremony – the university created a kind of diploma for the best weavers – and there is also a prize for the most beautiful carpet. Above all, these gatherings are important for their social aspect, the exchange, and the recognition from the university. And, of course, the women and men from the cooperatives always enjoy themselves very much, and so do we and our staff. Working for the DOBAG Project is different than working just anywhere. There's a certain amount of dedication involved.

In this university–cooperative business relationship have there been times where the different organizational structures of the partners might have given rise to misunderstandings or incomprehension? Have you had to make adjustments in order to continue working together?

HARALD BÖHMER: I used to go from village to village, now my colleague Atlihan does that. But actually, through the personal contact involved, we normally don't have any problems like that. We did once. It was at a time when everything was running smoothly, and a private company approached the villagers with an offer to buy the whole cooperative. It wanted to pay for an entire year's production in advance, but in return it wanted to choose the chairman and demanded that the cooperative cut off its ties to the university, arguing that the villagers had already learned how to dye their wool. Of course the plan was to gain control of the whole project including the success it had attained up to that point and to utilize the cooperative's expertise for its private-sector purposes. But the members of the cooperative unanimously voted against it. That made us very happy because that showed us we were doing a good job and that the people in the cooperative understood that the DOBAG Project guaranteed them independence. Since the university had always been fair about marketing the cooperative's rugs, and it was easy enough for it to raise the 4 percent of its sales it was required by contract to pay the university, the cooperative would certainly have worsened its situation by taking the offer.

If I understand correctly, the DOBAG Project is much more than other, more ordinary models for cooperation between an academic research institution and an "industrial" partner. After all, it wasn't as if two already existing "partners" simply came together; instead, you actually founded the cooperatives in western Anatolia. Do you think the cooperatives would have evolved so well without the university as their partner?

No, definitely not. Because we didn't just teach them about old dyes and dyeing methods; we also took charge of marketing the rugs and have consistently given the cooperatives advice on what buyers are looking for. For example, we always advise the women to produce only a few of the so-called bestsellers to keep from saturating the market. Their lack of language skills and education would have kept the villagers from being able to do any of this alone. And since from our side this cooperation isn't profit-oriented, since to us it is a cultural and development aid activity, it was a win-win situation for the cooperatives.

I doubt they could have been successful without us or with a profit-oriented partner. From the beginning, our focus was on ensuring sustainability and preserving old traditions. A private-sector partner probably wouldn't have pursued the research on natural dyes because, when I started out, it didn't look like it would ever turn into anything profitable. To me and everyone involved, it was more of a personal interest. And in the framework of the university as a public institution we had the ideal conditions for carrying out our project.

RENATE BÖHMER: Our primitive dye recipes have spread by word of mouth to other regions in Turkey, but that still hasn't made other cooperatives or villages successful, one reason being that they don't have the support to go with the know-how. They didn't follow through all the way, there were no quality-control inspections – of course not, since who would conduct them? That goes to show how important this partnership was and is. Without the protection of the university the cooperatives would have surely ended up at the mercy of private businesspeople, and against them they wouldn't have stood a chance.

Do you as a university pass your knowledge on to just these two cooperatives or to other people too?

HARALD BÖHMER: Others have tried to profit from us: dealers wanted to send us fiber samples, for example, and have us confirm that they also use natural dyes. They wanted us to issue them a certificate. But we can't do that because how do we know that the dealers really weave their rugs with the fibers they sent us? But anyway, we couldn't handle any more cooperation projects because the university doesn't have enough staff to supervise them.

And these rug-weaving areas are far away: one cooperative is located more than 400 km, the other 700 km from Istanbul. It's too much. You're on the road for weeks going from one village to the next, and then there's another aspect: the market isn't that big because compared to machine-made rugs that use synthetic dyes, which are much cheaper and easier to produce, our carpets have their price. If we started other similar cooperatives, this would hurt the existing cooperatives. And we don't want to do that.

How did you manage to implement your wool-dyeing know-how in the villages? That costs money. Where did it come from?

To us it was important to be able to pay for the rugs immediately. Two well-to-do friends of mine, both doctors, each gave us over DM 10,000.00 so we could pay for the first few carpets. We paid them back and later the cooperatives got a large interest-free government loan. This was paid back in a few years. Today, approximately 200,000.00 US dollars are distributed to the cooperatives annually.

How much money did you need to start your research at the university?

In the beginning we conducted our research in our kitchen, not at the university at all. Later, the university set up a laboratory for us using our German Ministry funding, a total of DM 30,000.00, and I was also receiving support from the German Research Foundation (DFG). Now all our research on natural dyes is financed by the cooperatives themselves, from the 4 percent of their sales.

Have the cooperatives ever come to the university with research requests? For example: We want to work with pink. Can you find out how we can do that with natural dyes?

Pink is a popular color for cotton socks, but it isn't a traditional color in carpets, so no, that has never been asked for – and certainly not by our buyers. The greatest demand is, of course, for colors that are lightfast. And for that reason we have had to alter some of our recipes. At first we were using onion skins for yellow, but then we noticed that the wool would lose its color within three weeks. So then we switched to dyer's rocket, a plant that is also native to Europe. But above all what the cooperatives expect of us is that we take care of the marketing for them. They want us to organize conventions, spread the word about them, and help market their product.

What were some of the mistakes made during the course of this cooperation project?

RENATE BÖHMER: According to the Cooperative Association Law, after deducting all expenses the profit is to be divided among the members. At first we paid the women according to what they delivered. But after a year, we changed things because we could tell it wasn't working. Now we calculate each woman's share of the profit based on the knot count of only those carpets sold within two business years. We no longer go by square meters because if you use thick wool, you finish a square meter faster than if you use thin wool. Paying by knot is fairer. And we only paid for carpets sold within two years. If a rug isn't sold in that time, you get nothing for it. But I must add that every woman initially receives a base sum in cash for each carpet she completes. This covers her expenses and she has earned a bit on top of that. This system has been working well for more than 20 years.

Know-how, self-empowerment: regional development

What would be your advice to people who are already involved in or want to start this type of cooperation project?

HARALD BÖHMER: What is important is that you are always there and you can demonstrate things to the people in the cooperatives. It's important to not just hand out recipes but to also show exactly how – in our case – dyeing is done. If you want to teach something, you have to be able to demonstrate it too. We couldn't have taught the women how to weave, but of course they didn't need us to. They needed to learn how to dye, and we demonstrated that to them, again and again. We had taught ourselves everything we knew, had read a lot of literature on the subject, followed a lot of paths that led nowhere, for example saffron. You find it everywhere in old recipes and books, but in fact it was never used for carpets.

*Saffron was expensive and its color didn't hold. That was something we found out through experimentation too. But the main thing is that you have to learn the language and culture and be open to life in the villages. That is an essential prerequisite. And whatever you do, don't go about it as if you were running a business, that doesn't work. You have to set up very professional structures, and it's extremely important to be very reliable and consistent with your partners. For example, the members in our cooperatives must deliver high-quality products. That is something we demand. And in return, we are dependable partners who aren't going to abandon the villages at the drop of a hat. **In the villages themselves you need people with strong personalities who take command and look out for the best interests of the cooperative.** Good work on both sides and mutual trust, that's the bottom line.*

Now after more than 20 years, could these two cooperatives get along on their own without Marmara University?

RENATE BÖHMER: No, not yet, and that's because of marketing. The people there still can't maintain their international contacts. Besides, the global market doesn't stand still and tastes change. Soon, young people in western countries may no longer be interested in oriental carpets with traditional patterns. You have to keep these aspects in mind, keep your eyes and ears open, and advise the cooperatives accordingly. There are already quite a few staff members at the university who are very good at this. They watch the market and determine what people want. In this way our dealers are also an important source of information, as is reading international journals and actively taking part in the scientific discourse in this field.

If I understand you correctly, not only the villages involved in the cooperatives but also the surrounding regions have profited enormously from the DOBAG Project.

HARALD BÖHMER: Yes, the entire region! Not just the participating villages, but the other ones have also suddenly taken a giant step toward "modernization." Modernization, relatively speaking, because here we're talking about things like introducing electricity to a village. When we started out, these places had neither electricity nor telephones. Now everything has changed. People from the surrounding villages have moved to the villages that belong to the cooperatives because they can send their children to better schools and escape the social restrictions of their native villages. They don't move to the slums of Istanbul anymore. Today the people here have their televisions, their tractors – often jointly owned by several families – their mopeds. And their living standards have improved enormously. In agricultural regions, people have bought land – a secure investment. And when it's time to harvest the olives, they can give local people work.

However, in the region around Ayvacik there is a new challenge on the horizon: an American company has settled nearby and plans to cultivate genetically modified corn. And they pay higher wages. The prices for our carpets have stayed the same for 25 years. The reason for this is that, one, we are dealing with a small market and, two, we have to compete with cheap rugs from China and Nepal. The question is how long the Ayvacik cooperative will survive. Young women are constantly losing interest in this old craft, but the key aspect in this new development is the fact that the American company pays more.

In the Yuntdag region things are different. There, there is no alternative line of work; on the contrary, more and more women want to join the cooperative. In this region the DOBAG Project definitely has a future. Moreover, over the years we have built up a strong network of customers that ensures that in the future there will still be a demand for the high-quality product being made by the cooperatives.

Thank you for the interview.

Key project insight

Important is …

- a willingness to make a long-term commitment.
- that the research results may have come from the German partners, but that the task of imparting this knowledge and checking up on the villages is done in close cooperation with native staff.

- self-empowerment and the establishment of self-governing, decentralized structures.
- a structurally anchored strengthening of the women who produce the carpets within the traditionally male-dominated society in which they live.
- the establishment and ongoing maintenance of a complete business process: research, instruction, organizational development, quality management, marketing, and distribution.
- that the foreign partners have a profound knowledge of the culture and language of the country.
- that there are clear contractual agreements, consistency in respect to quality and economic efficiency.
- ongoing personal, face-to-face supervision.

The Ease and Burden of Starting Over: Research Cooperation After Communism

Interview by *Mathias Bölinger*

Slovenia

ALNIŠKO VODENJE Z HTEVNEJŠIH ROCESOV
KEMIJSKI IN PROCE SNIH INDUSTI JAH
Projekt: RACIONALIZACIJA PROCESOV ZGOREVANJA
ALNIŠKO VODENJE Z HTEVNEJŠIH ROCESOV
KEMIJSKI IN PROCE SNIH INDUSTI JAH
Projekt: RACIONALIZACIJA PROCESOV ZGOREVANJA

"The problem facing us is that we don't have a proper business mentality. That doesn't apply just to Slovenia but to western Europe in general. Students in Europe don't ask themselves how they can use their skills to become successful in the business world, they just want to find a job."

From the atomic bomb to the promotion of regional development: the Jožef Stefan Institute in Ljubljana integrates international academics into Slovenian companies and Slovenian companies into international research projects.

A company from the steel sector and Slovenia's renowned research institute left Yugoslavia behind when they set out for Europe. Have they reached their destination? Or have they moved on? Maybe they want more than the complaisant comfort Europe has to offer.

An interview on small enterprises and big potential. On Slovenian high-tech products and the global economy. And who has to talk to whom about what for both to be possible in Ljubljana.

Founded after the war as one of three centers designated to conduct Yugoslavia's atomic energy research, the Jožef Stefan Institute (IJS) is today the most important non-university research institute in Slovenia. Since the sixties, the original field of research has been expanded to include numerous high-tech disciplines. Today, in addition to nuclear physics IJS' nearly 500 scientists conduct research in the fields of molecular biology, biochemistry, chemistry, nanotechnology, and information technology. Its budget, which was 35 million euros in 2005, is financed largely by the Ministry of Education. With the aim of attracting technology companies to the area around IJS, the Institute co-founded the Technology Park Ljubljana in cooperation with major Slovenian technology companies.

Kogast is a corporate group that manufactures gastronomy equipment for restaurants, hospitals, and cafeterias. Its clients range from Bulgarian Black Sea resorts to hospitals in Sarajevo or restaurants in Switzerland. Kogast is a member of the Network of Manufacturers of High-Tech Products.

Mitja Jermol, MSc, is head of the Department of Information Technology at the Jožef Stefan Institute and founder of the software company 3K IT, one of the tenant companies at the Technology Park Ljubljana.

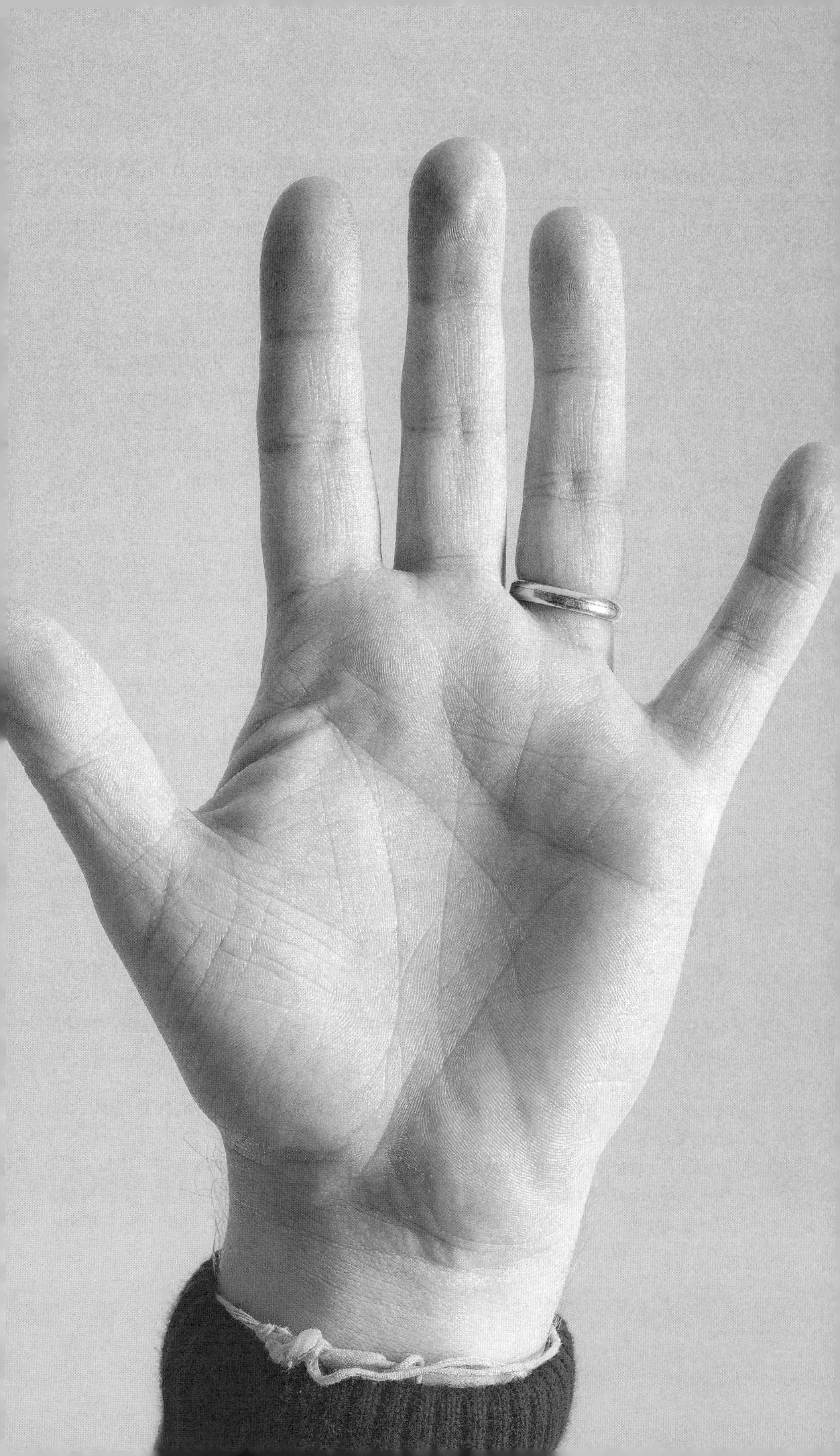

Marko Jurančič, Dipl.-Ing., works for a Kogast subsidiary and is coordinator of the Network of Manufacturers of High-Tech Products.

Mr. Jurančič, Mr. Jermol, for a year now you have been working together on a software project. How did this collaboration come about?

MARKO JURANČIČ/KOGAST: Mitja Jermol was involved in organizing the project and asked us if we wanted to join. He was looking for someone who would be involved in the project as a company, i.e. would provide data and share experience, and who was also part of a network of interconnected companies.

MITJA JERMOL/IJS: Essentially it is a joint platform of different European institutes and organizations who have got together to develop a software tool. The list of participants includes German, Italian, Hungarian, and Polish organizations and business associations. The explicit goal of this project is knowledge transfer from western European business associations to their eastern European counterparts. That's why it was clear to me that in addition to us there had to be another Slovenian representative, and if possible a company that was also part of a network so that it could pass on the new knowledge and experience.

MARKO JURANČIČ/KOGAST: Roughly speaking, our objective is to develop a software tool that will give the companies a better overview of the production capacities of the other companies and the products they offer. Among other things, this information would facilitate the calculation of production times.

Something like a European database with company coordinates?

MITJA JERMOL/IJS: Technically, it's more than that. What we want to develop is software that helps the organizations consolidate their databases. To do this, we have to make different systems compatible, and that on a European level. We're bringing together different companies in different places who speak different languages and have different technical standards. We want to integrate the different applications that exist in different business networks so that what we come out with in the end is something that resembles a supermarket: an assortment of applications intended to facilitate international cooperation among companies.

That means IJS helps network Slovenian companies across national borders.

That's one of the Institute's major objectives. We want to make it possible for Slovenian companies to participate in international scientific developments. That applies not only in my area of expertise – network technology – but also in physics, chemistry, nanotechnology, and biochemistry.

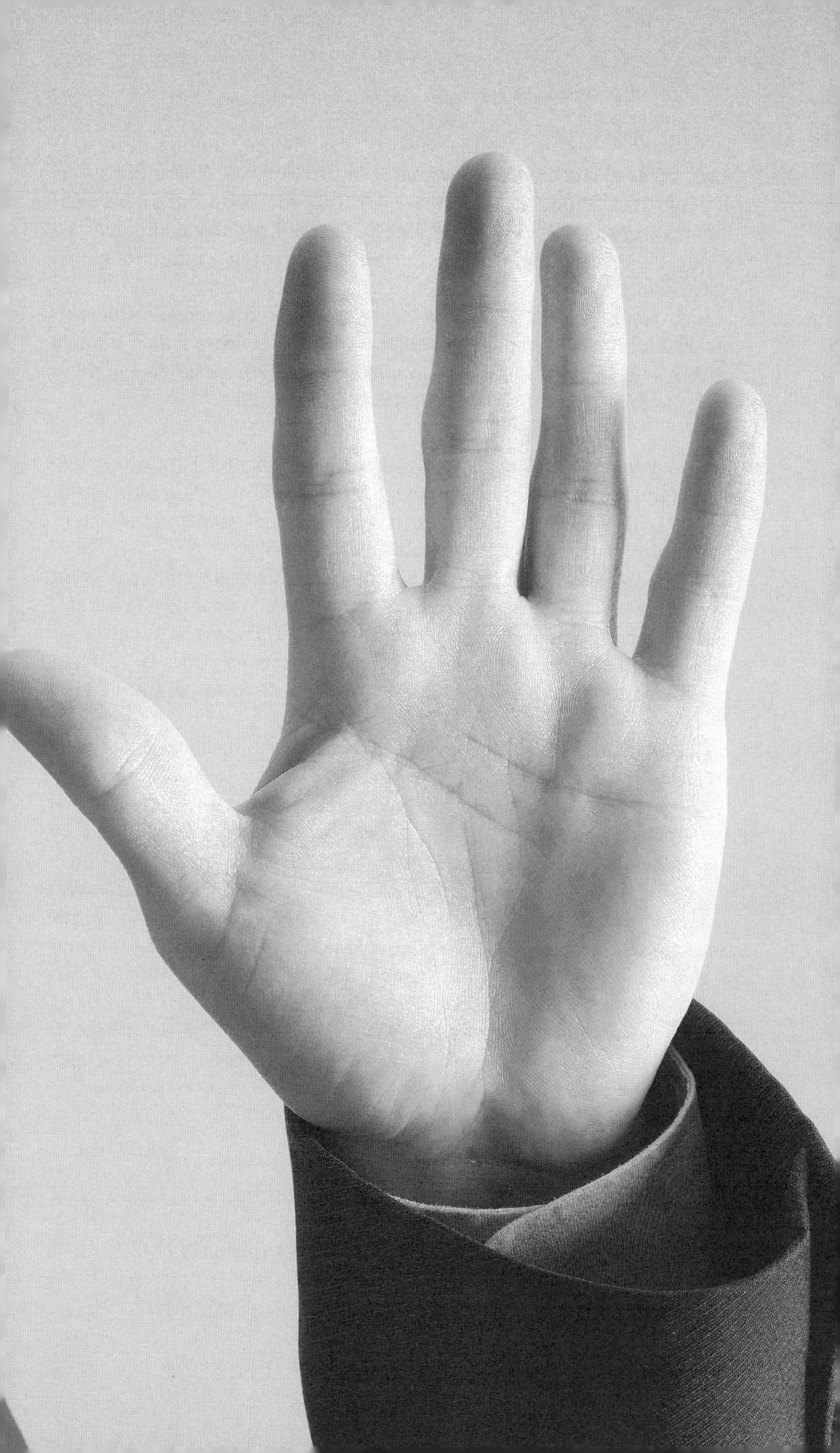

Mr. Jurančič, you say your company was chosen, among other reasons, because you are the coordinator of the Network of Manufacturers of High-Tech Products. Before we go into the cooperation between the Institute and industry, I'd like to know how the companies go about organizing themselves – especially when it comes to global networking.

MARKO JURANČIČ/KOGAST: It is, of course, easier for us to connect with other research facilities or companies in other countries as a network than it would be as a single company. It is quite common here for companies to join forces and start joint cooperation projects with IJS or universities.

The Slovenian economy, and I assume also the metal industry, had strong ties to companies in the other former republics of Yugoslavia prior to 1991. To what extent did you have to reorient and to what extent reorganize yourself after the dissolution of Yugoslavia?

Overall, Slovenia was lucky. Our industry wasn't oriented on the needs of the army as much as those in other constituent republics. We always concentrated more on exporting to the European market and the Third World. As part of the Non-Aligned Movement, Yugoslavia focused its economic policy on countries like India or various African states, and we've held on to these contacts to this day.

But allow me to elaborate using our company as an example. We build commercial kitchens. We used to produce a lot for the Yugoslavian army. Now we export mainly to the European Union, Russia, and the Arab Gulf states, but with this new orientation we are also confronted with new demands that we cannot handle alone. One example from my field: today we are expected to offer kitchens that are partially or entirely computer controlled. Those are common customer requests. This means we have to work with software providers if we want to stay competitive. And now the problem we are faced with as a company is that the software I employ in my commercial kitchen has to control not only the components that I manufacture myself but everything else I buy as well. The companies involved have to develop a common standard. In order to do this we had to organize in some way. The central planning offices we used to have don't exist anymore, so we have found a way to work together on our own.

Photo   (**see page 249**) Mitja Jermol, head of the Department of Information Technology at the Jožef Stefan Institute

Research and education as a state-financed service for small and medium-sized enterprises

Mr. Jermol, we were talking about how these networks also play an important role for you as a contact person. On the other hand these structures are still being set up. How realistic is it for small enterprises – especially those that might not have a research department with an adequate budget and the contacts with academia – to be able to collaborate with you?

MITJA JERMOL/IJS: We are still being financed largely by the state. We receive part of our budget directly for basic research, but the other part is reserved for projects. That's the part that is interesting for small enterprises. The deal is this: a company comes to us with a problem. It can then apply to the government for resources, which we receive to use for research. The company isn't required to pay anything; funding for the researcher's work at the Institute comes entirely from the government. This is the model that has been in place here in Slovenia since the eighties: the government makes researchers available to the companies.

If I understand correctly, this money is earmarked for cooperation projects. If no one submits a project, you don't get the money.

Yes, that's right, for industry projects, i.e. projects in which the Institute is called upon to solve a specific problem. Technically, this is how it works: if a company is allotted the research money it applied to the government for, it is required to spend at least 50 percent of it on us. It's the government's way of making sure that we produce results instead of just sitting around. The budgets in question, however, are relatively small.

You mentioned that this principle was introduced in the eighties. That means the Institute was already a scientific service provider for companies back in the days of Yugoslavia?

After the war, IJS was first founded as one of three research centers for the Yugoslavian nuclear science program. That also explains why one of our main fields of focus to this day is physics and nuclear engineering. In the sixties the Institute redefined its mission. Its new objective was the transfer of knowledge from research to industry. Unlike with universities whose aim it was to combine knowledge and teaching, the objective of our institute was always applied research, and this also included cooperation with companies.

Photo   (**see page 253**) Marko Jurančič, Coordinator of the High-Tech Network

MARKO JURANČIČ/KOGAST: Particularly for very small companies this model is often the only way to gain access to new scientific developments.

MITJA JERMOL/IJS: And this continues to be very important for our work. Marko mentioned previously that our industry structure consists of predominantly medium-sized businesses. According to European standards almost all Slovenian companies are small or medium-sized enterprises, the definition being enterprises with fewer than 250 employees. In a country with 2 million inhabitants you can imagine how a company with 250 employees might be considered large. So in Slovenia, when we speak of small companies, we mean companies with 10 or 20 employees. And we need to offer these enterprises special advanced training programs. Large companies want a program especially designed to meet their needs and are willing to pay for it. Small companies can't afford this. That's why we offer a wide range of seminars and lectures open to all interested companies, and we have started recording these lectures and making them available on video: lectures given at our institute, at the University of Ljubljana, and other universities around the world. Anyone interested in a specific topic can download the lecture from the Internet and listen to it. **We already have 1,300 of these lectures available online, the largest database of video lectures in the world.**

You mainly mentioned the services that IJS makes available to the companies free of charge. What about specific research cooperation projects themselves? What percentage of your cooperation partners are small enterprises?

Some cooperation projects fall into the government promotion programs I mentioned earlier. But most of our cooperation partners are larger companies, companies that compete in the global market. They know that they require very sophisticated expertise, and for that they need access to international know-how and high technology.

In this context you have turned your attention to training and educating new generations of scientists for industry. You founded a postgraduate program in which cooperation with industry is a requirement for participation.

The Institute set up a postgraduate school two years ago. The program is international – instructors and students come from different countries. We founded this school in collaboration with the most important Slovenian industrial sectors. The core idea of the program is to foster close cooperation between the school, the Institute, and industry. Our aim was to provide hands-on training.

That means that thesis and dissertation topics all address real scenarios, real problems that specific industry sectors need to solve. On behalf of the company, the student takes this problem to the school, where there is a committee of professors that decides how this problem can be turned into a dissertation or thesis topic. The idea is that in the end the student will have got his education and degree and at the same time he will have solved a concrete problem for his company. And in the process he will also have made important contacts within the Institute. This year was our second year and we had 60 students. Our first year we had 45. It's a research internship where the student gains work experience and studies at the same time, and the company is responsible for paying the student's tuition.

That means the student's training is covered mainly by third-party funding. What is the research situation in general, what percentage of IJS' funding is put up by a third party?

We hope that in the future we will be able to increase third-party funding for our research. At the moment we raise approximately one-third of our budget from companies and other promotion funds, e.g. EU programs; the state pays the rest. This entails enormous restrictions, e.g. we don't have the right to do business with our research. We're not allowed to use it to make a profit. That means the Institute can only make very limited investments. We can't start our own companies. It is theoretically possible to apply for a permit, but it's unrealistic because it entails going through a long procedure for each project. This is one of the major problems, and we're working on changing it.

The invention of entrepreneurship

You took things into your own hands and started your own company…

Some of my colleagues and I realized that there were certain tasks we couldn't tackle at the Institute and since we couldn't let other companies take care of them either, we decided to start a small company of our own. We developed software that lets you systematically analyze client and business data. But what was more important to us was to show what opportunities there are in starting a spin-off.

As an IJS staff member, are you even allowed to start your own company?

Yes, no problem. You don't even need a permit.

What about sideline work, are you allowed to work for other companies?

At the moment, that's still a gray zone, it's a legal loophole. What we do have is a consensus that it needs to be possible. We currently still have to apply to the Executive Board, but it's not clear what they base their judgment on.

All that is going to change soon. We're working on new regulations for sideline occupations.

Then you agree that even IJS as a research facility should take a more business-minded approach?

We're faced with the choice: either we become a pure research facility or we do meaningful work in cooperation with our environment. The problem facing us is that we don't have a proper business mentality. That doesn't apply just to Slovenia but to western Europe in general. Students in Europe don't ask themselves how they can use their skills to become successful in the business world, they just want to find a job. Unlike in the USA, European PhD students rarely ask themselves if they will be able to market their idea. That's the main problem. In Slovenia, if someone is successful, we don't say: Look at him, he made it! We ask ourselves: Where in the world did he get the money? It doesn't occur to anybody that someone with commercial success might have had a terrific idea and some good business sense. That's the way it is in Slovenia, and that's the way it is throughout Europe. If this mentality doesn't change, every reform IJS or the university tries to push through will fail.

Mr. Jurančič, you are also confronted with the same issues…

MARKO JURANČIČ/KOGAST: Yes, that is exactly how I see it. We aren't innovative enough, and don't take advantage of our potential. When we started to build our network, we had big problems getting some of the companies to get their stuff together and be reliable, meet deadlines. They just went on doing things the way they always had. Try to coordinate a joint project under those circumstances! We had to deal with the principles of the organization before we could even create the foundation for joint projects. This is slowly starting to change – on the political level too. Until only a few years ago, economic promotion in Slovenia consisted of supporting business sectors that weren't particularly lucrative, e.g. the shoe or textile industry. Today, funding is oriented more on product marketability.

MITJA JERMOL/IJS: Despite the complaints, you have to admit that there are big changes taking place in the minds of entrepreneurs and scientists alike. Young people are entering the business world. They didn't grow up in the old system, and they have an above-average rate of success. The second aspect is that the Slovenian companies have grasped the fact that they have to keep up with the global competition. Before, it used to be enough to be the best in Slovenia. Today they have to be the best in the world. That has temporarily destroyed the image that many companies have of themselves, but the pressure is making us reorient ourselves and it also forces us to work with other partners. That's very, very positive. What is less positive, as I see it, is that the governments – in Slovenia as well as other European countries – still haven't got that through their heads. At institutes and in companies the willingness to work together is much greater than the government's willingness to create the necessary framework.

Is it more difficult for the Institute as a state facility than it is for the companies?

We are an old institution, and we are and have always been a rigid system, but it's getting better. And compared with the university, we, as a pure research institute, are better off. Universities have their own economy, their own closed circuit, and are in a certain sense closed institutions. IJS is at least active in a market environment. We have to compete with institutions all over the world. We have to be more flexible. I've been working there for five years and a lot has changed during that time.

In your experience, what are the typical points of conflict between academics and companies or businesspeople?

Typical communication difficulties begin with how the people involved perceive the project's objective. Academics don't have much of an understanding of business interests or business management issues, and companies and businesspeople, for their part, often expect research labs to function like production plants. I would say that there are considerable differences in the basic motives and mentalities.

Do you have special strategies for tackling these types of difficulties?

We don't have any one strategy to solve all these problems. But we have, of course, learned from our mistakes. Our most important lesson has been that you need to communicate, and I mean a lot. We solve these problems on a case-by-case basis, and we've learned that with each of these cooperation projects we have to invest a lot of time to understand what our mutual motives and goals are. Then things usually run more smoothly. But of course this doesn't always work. There have been projects we've had to cancel right at the start. Mutual trust is very important.

What we've also tried, and what seems to be a good solution, is calling in facilitators. That cuts down the time needed to understand the different perspectives. There are people who have worked in both environments and who are thus familiar with the problems and motives of both sides. These are the people who can help in these kinds of situations.

The technology park as an incubator

You have developed very concrete strategies to bring the business world and IJS together spatially and institutionally. I am thinking specifically of the Technology Park Ljubljana, which you founded in conjunction with the city and seven large companies. Who is allowed to settle in the Park?

A company that wants to settle at your technology park is required to submit a fully drafted business plan to the approval commission, which is made up of businesspeople, academics, and representatives of financial institutions. The conditions are that the business idea is innovative and has to do with technology. Moreover, it must also have promising prospects. **The companies usually stay at the technology park for three years, after that they should be viable on the market.** *Currently we have 67 companies, most of them involved in the fields of information technology, communication, biotechnology, new materials, automation systems, and environmental technology.*

Which conditions and types of initial aid does the technology park offer its tenants?

Most importantly, the Park offers a professional infrastructure and helps finance administration costs. The companies receive municipal subsidies to cover things like telephone, Internet, electricity, and lease. There are also subsidies for certain business administration tasks like bookkeeping. The companies can take advanced training courses in business management subjects, and we also offer a range of information on new business opportunities. The technology park promotes networking among the companies and offers basic legal aid. The city actively supports the growth of the technology park and is currently building new facilities at a second location.

What role do IJS and the tenant companies play in cultivating young entrepreneurs at the technology park?

The Institute helps make good ideas or prototypes competitive and at the same time provides intellectual capital, supporting the companies in different research issues. But normally we do not get directly involved with new companies. The tenant companies are not very active. Some might invest in start-ups, but not usually.

To you, what is the most convincing reason for tenant companies in the technology field to settle here?

The technology park lies in the middle of the Slovenian research landscape. In addition to IJS, there are other technology companies and the University of Ljubljana located nearby. Besides, infrastructure costs for companies in

Ljubljana are relatively high. The municipal subsidies here are no insignificant factor, especially for small companies.

East research, west research?

You have referred many times to the fact that the problems found in Slovenia are the same as those facing other European countries. Do scientists these days see being in the east or the west as a relevant issue?

*Definitely. That's deeply ingrained in us. **As an eastern European scientist you are still seen as someone from behind the Iron Curtain. Even if you are doing international cutting-edge research.** In the Department of Knowledge Technologies, which is where I work, we have grown enormously in the past few years and I dare say that in our field we are the leading institute in Europe. But when I approach companies in the west, this hardly makes an impact. As an eastern European you have to be ten-times better to be perceived as equal. Our only chance, therefore, is to concentrate on our neighbors. People know us in Graz or Vienna: there we have less of a problem.*

What comparative advantages do eastern and central European countries have?

Our mentality is certainly different than in western Europe. The new member states are more creative and open. The energy to really work towards the big breakthrough is something you find in the former Eastern Bloc countries. If you go to western Ukraine, you will notice a veritable boom of start-up IT companies. You will find lots of very promising software developers. The same goes for Bulgaria or Lithuania. In western Europe there is no comparable accumulation of new technology sectors. Why not? Because in these countries in transition motivation is higher. People here are used to and open to big changes. The question is how long will we be able to hold on to this?

Which changes can we anticipate at IJS, what can we expect in the near future?

In the future, we want to invest more in industry-motivated research. We're in the process of adjusting our structures and regulations so that we can do more commercially oriented work.

One point I am really pushing is the right to be able to start spin-offs. We are currently looking very closely at models in use at other research facilities. We will implement parts of these models, perhaps not one to one, but the main parts. We are trying to find new ways to shift our work priorities more towards putting our research results into practice. What we want is to become the leading institute in Europe in the fields we work in. Like all other eastern

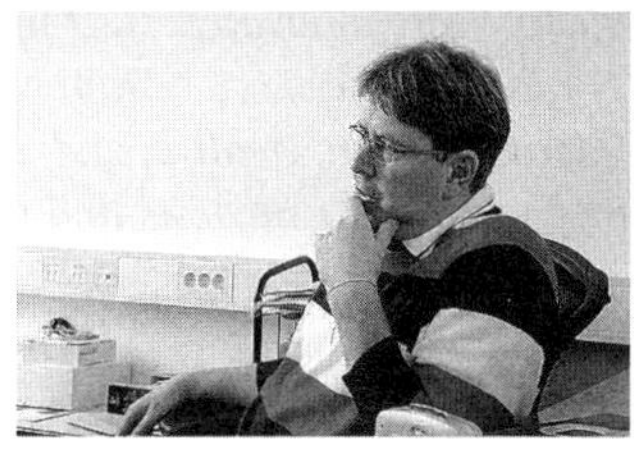

European countries, Slovenia has suffered from the brain drain. Our best minds went abroad. We want to make sure they can find the conditions necessary to conduct excellent research here. We want them to come back. And for that, in addition to outstanding scientists we also need to attract the companies those scientists find interesting. This is why we founded the technology park. And last but not least, the third thing we need is a good network to make us attractive as an international research site. These are the things we are currently working on.

Mr. Jermol, the strategies and examples we have been discussing cover widely ranging fields. From research and education to financial incentives and service providers. From concrete infrastructure issues to communication strategies. Summing things up, what can other research institutions and companies learn from IJS?

The most important lesson we've learned is that no research institution can survive globally if it doesn't take the initiative to do market-oriented work and set up a good network. We have shown that even small countries like Slovenia can be the best in the world in certain research fields and that we can also nurture very innovative start-ups that go on to become market leaders.

What you need is a certain winner mentality and intellectual flexibility. You have to be open, creative, and willing to cooperate. And you need a stimulating mix of business interests and a branch of research that isn't under too much pressure.

Thank you for the interview.

Key project insight

With a traditionally strong orientation in applied research, the Jožef Stefan Institute has in the past few years witnessed a veritable boom particularly in the new sciences, e.g. knowledge technologies. Through all this, the Institute continues to make industry and regional promotion its top priorities. Since the eighties, an important strategy for promoting the economy has been to offer research services free of charge – a model that currently gives especially very small companies access to international-level technology.

Now that Slovenian companies not only have a stronger position vis-à-vis the global competition, but are also part of a well-developed network, they also have access to new cooperation models, e.g. cooperation projects on a European level. This includes PhD and master's programs in which students are only accepted under the condition that they submit their research topic in collaboration with a company. One of IJS' main objectives is to offer companies a way to connect with international-level research, and it augments its program range with services like its database of videotaped lectures. In the future, researchers at IJS want to increase their own business activity. The legal framework for this is still being drawn up.

Slovenian companies have begun to connect among themselves and build networks. In this way they are filling organizational voids that have resulted from the transition from a planned to a capitalist economic system. These networks allow companies to coordinate technological developments among themselves. In the Network of Manufacturers of High-Tech Products, for example, members seek to develop shared software standards. At the same time, these networks provide their members with contact people for transborder projects – for IJS too – and also serve as a knowledge-transfer multiplier.

Generalists Who Can Specialize: The Cultural Scientist as the Better Economist?

Interview by *Martin Rasper*

Germany

Gesellsc
sind
Wir a
Zep

e Probleme
pliniert.
niversity
Sportgelände
PSG

In der Geschichte dieses Planeten ist der Mensch seit den letzten neuntausend Jahren mit wissenschaftlicher
Arbeit beschäftigt gewesen. Mit welchem Ergebnis. Er kann Unkrautvernichter herstellen, aber kein Unkraut.
[George Spencer-Brown]

"Structurally, the question is how do we manage to get decentralized intelligence to lead to a centrally intelligent decision. And that for example means taking approaches like integrating the most ignored body of people at universities: the students. That's my only advice to state-run universities: Talk to your students and alumni! They are the ones who experience the university as a whole."

Founded in 2003, the private Zeppelin University in Friedrichshafen on Lake Constance is the youngest and currently perhaps most innovative university model in Germany. With its radical interdisciplinary approach, a consistently international orientation, and – not to be underestimated – the charisma of its dynamic and youthful president, Stephan A. Jansen, it cuts an exotic figure in the German university landscape.

In large letters on the façade of the gabled building that used to house a flight school in the thirties are the words: "Social problems are unruly. So are we!"

The "most daring start-up university in recent years" (*Süddeutsche Zeitung*) – made possible by the substantial financial support of the Friedrichshafen companies Zeppelin GmbH, ZF Friedrichshafen AG, and Weishaupt, specialists in heating technology – has set some unusual principles for itself: strict interdisciplinarity, international orientation, and a customizable curriculum that draws strongly on the involvement and responsibility of its students.

"We don't tell our students what they are supposed to learn," explains University President Stephan A. Jansen, "we tell them: Here you have the chance to become who you are."

The three departments: Corporate Management & Economics, Communication & Cultural Management, and Public Management & Governance each offers accredited courses of study with bachelor's and master's degree programs based on the three-cycle system set down by the Bologna Process. No more than 30 students are accepted each semester. Annual tuition fees for the three-year bachelor's program are approximately 8,000.00 euros, and for the two-year master's program between 8,000.00 and 10,000.00 euros.

Columbus Holding AG in Ravensburg is a medium-sized holding company with an IT and media range that has made a name for itself in many sectors through its art-sponsoring activity. The four companies – Columbus Leasing, Columbus Office, Columbus Systems, and Columbus Interactive – each cover their business areas of expertise.

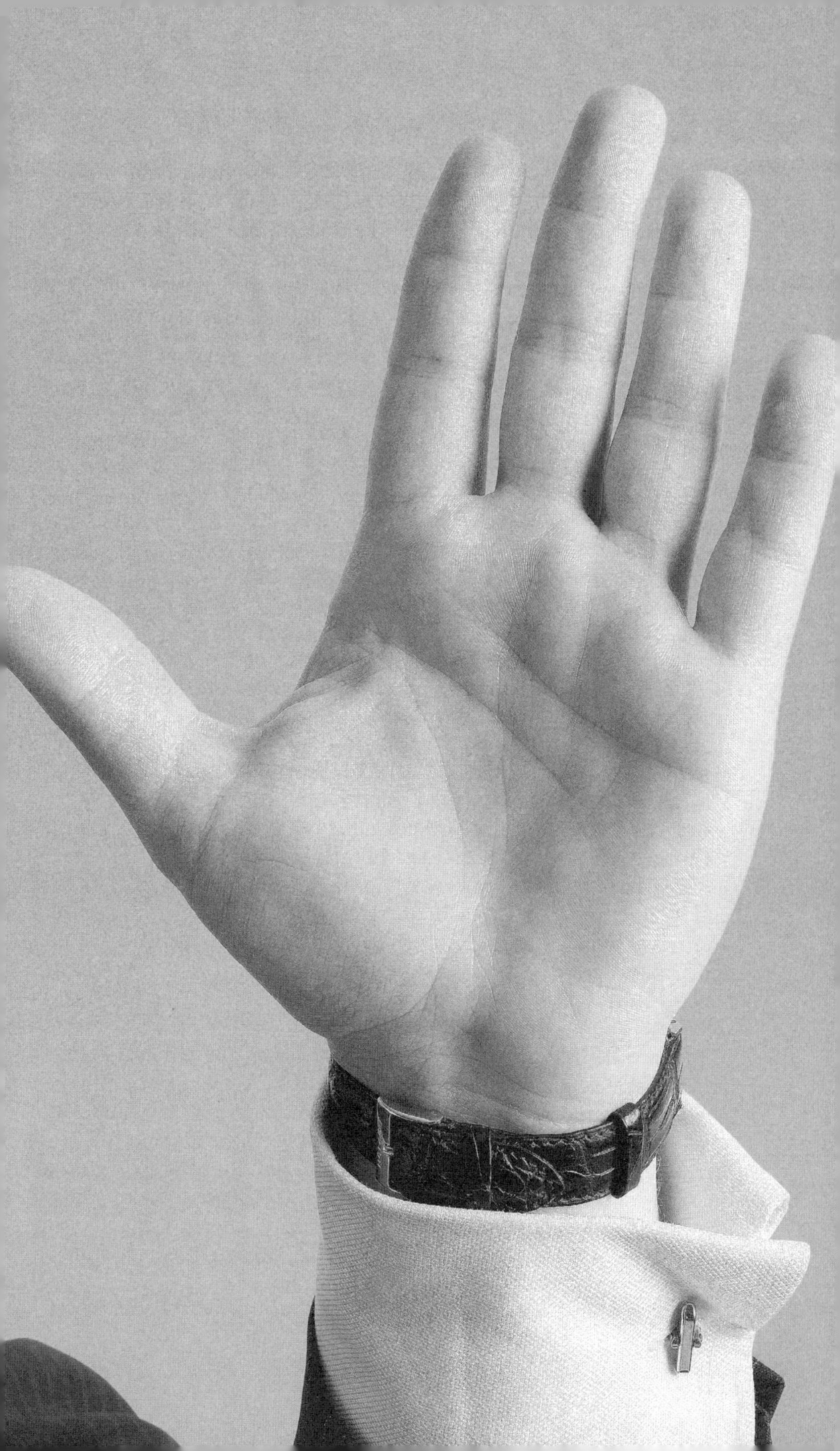

The Columbus Art Foundation, which also belongs to Columbus Holding, stands out for its innovative and effective promotion of young artists. To holding company president Götz-Wolf Wagener the company's commitment to art is part of its attitude and the view that "business activity doesn't take place in a social vacuum."

Together, Columbus Holding AG and Zeppelin University cultivate cooperation projects on different levels, in cultural management, for example.

Professor Dr. Stephan A. Jansen has always been the youngest at every step up the career ladder. After completing a bank apprenticeship and while still a student at the privately run Witten/Herdecke University, Stephan A. Jansen, born in 1971, acquired the capital to found his own institute: the Institute for Mergers & Acquisitions. Just a few years later – after sojourns abroad as a visiting researcher at Stanford University and as a visiting professor at Harvard Business School – he was asked to become president of Zeppelin University, which was then in the process of being founded. He accepted this post as well as that of the academic chair of Strategic Organization and Financing. In *Oscillodox* he describes how companies can oscillate between paradoxes rather than trying to get rid of them; it received numerous awards and was named business book of the year in 2000. His book *Mergers & Acquisitions* is on its way to becoming a standard work and is already in its fifth printing.

Götz-Wolf Wagener, born in Swabia in 1949, is a textbook example of a self-made entrepreneur: after finishing an apprenticeship as a waiter, he went back to school to study business administration. In 1992 he and two partners founded Columbus Leasing which has since multiplied and grown into Columbus Holding AG. His daughter's experiences as an independent artist have inspired his commitment to promoting young artists, which earned him the German Prize for Cultural Sponsoring in 2006.

To reach Zeppelin University by car you get off the B31 just before Friedrichshafen and turn onto a little street that takes you past a railway crossing. When one visits the university, the chances are the gates will go down just as one gets to the tracks, and one will have to wait a full five minutes for the train to pass. Thus the visitor finds himself apologizing for being late, but at the same time the incident is a welcome opportunity for Zeppelin University's president, Stephan Jansen, to launch into his favorite topic: the advantages of flexible, decentralized structures.

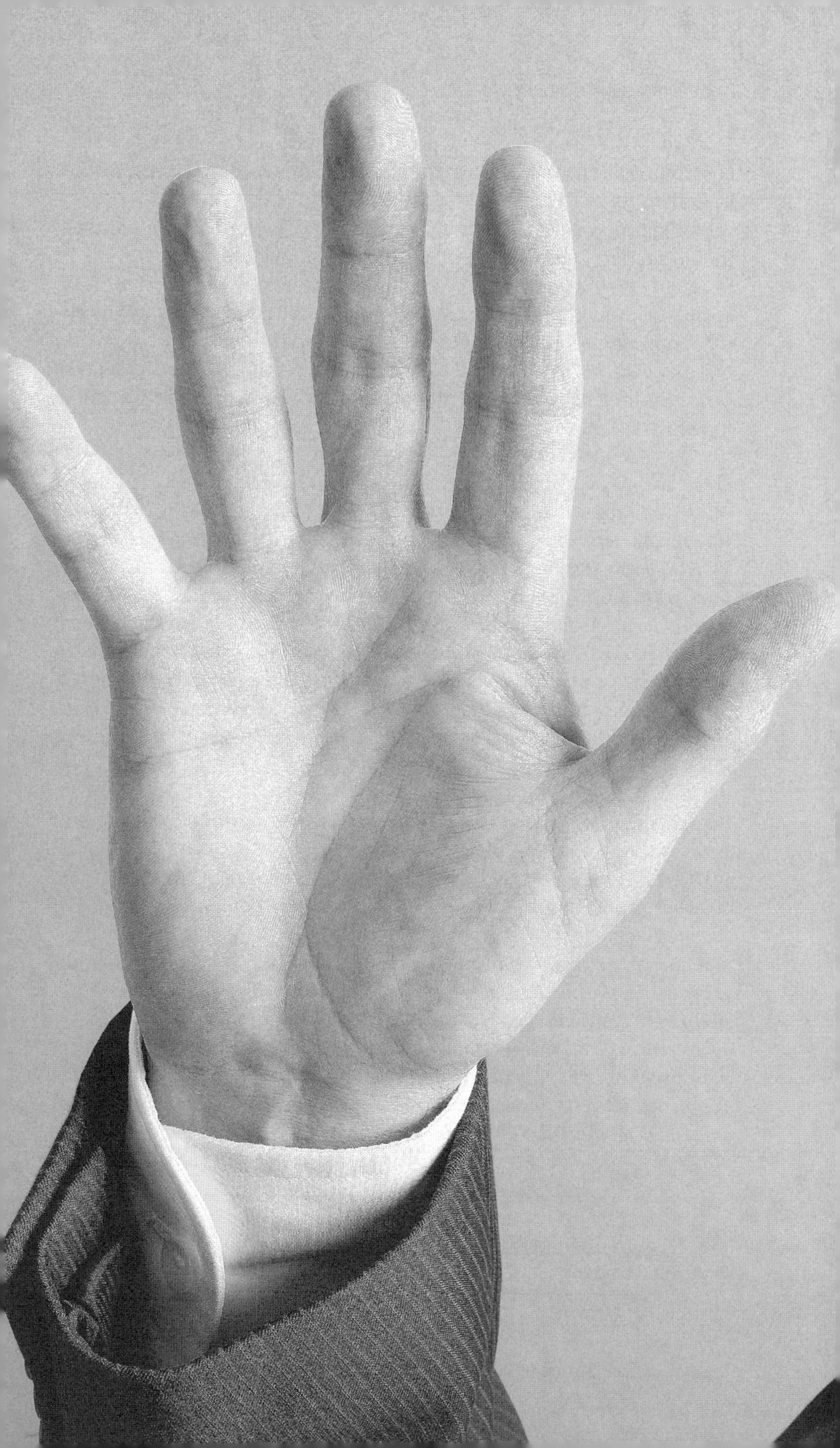

The university didn't have the problem with the gate until they started controlling all the railroad switches from a centralized dispatcher's office in Karlsruhe.

By the time we get started, my second interviewee still hasn't arrived, but no matter; President Jansen is eloquent enough for two.

Professor Jansen, people commonly view the cultural sciences and industry as opposites – on the one side, aesthetics; on the other, the hard world of capitalism. You, however, claim that cultural scientists don't just make good economists; they make the better economists. How do you come to this conclusion?

STEPHAN A. JANSEN/ZEPPELIN UNIVERSITY: That's correct, we believe that an expert with a background in cultural theory and theory of the state is better equipped to run a business than a manager with business knowledge alone. We are discovering that current management positions have reached their limits. The average staying power of high-level managers has dropped to less than three years for performance reasons. These leaders have serious communication problems. And after all, management is essentially communication. What does a CEO do? There are two comprehensive academic studies on the subject – from the 1970s and 1990s – that basically arrive at the same conclusion: 90 percent of their work consists of communication – conferences, phone calls, e-mail, meetings with employees, with customers, with shareholders, with labor unions, and the media. And less than 10 percent is desk work.

Even if a manager spends most of his time with communication, he still has to know something about business.

Of course, but the big problems are starting to blur, they don't conform to the bounds of scientific disciplines. That's why we need a new generation of managers who are not just experts in their fields but who also develop an understanding for neighboring areas. For example a museum director who isn't just a knowledgeable art historian but is also familiar with the cultural–political decision-making processes and sensitized to controlling and marketing issues. Or a business leader with training in media sciences and intercultural studies who therefore has a completely different take on the international and often conflicting interests of the shareholders, employees, clients, and other parties involved. Or authorities in the public sector who are capable of successfully

---

Photo   (**see page 281**) Stephan A. Jansen, Founding President and head of Zeppelin University

implementing modernization strategies using their communication and cultural skills, but who use business management instruments instead. This is our understanding of interdisciplinarity, and it is why we place such an emphasis on communication studies.

Interdisciplinarity is often just a catchphrase, used superficially without any real meaning.

Interdisciplinarity must mean that one is an expert in both fields rather than that one isn't any good at either one, which unfortunately is too often the case. What's important to me is that our cultural scientists are good enough to be able to end up in the business world. And they do – in practice as well as in academics. For example, we have a cultural studies graduate who just got accepted by the London School of Economics. And all nine graduates from our first year have found positions – with cultural institutions, but also with "purely" commercial businesses. **What we aim for are generalists who can specialize. And that works because each person is his or her own character – that comes from our university's focus on individualization.** *Our human-resources directors agree.*

Academic and practical-experience coaches

The principles of Zeppelin University sound impressive, but also a little lofty: you want to "develop individual passions instead of teaching a conventional curriculum"; instead of "lectures you have dialogue" and instead of ordinary counseling appointments you have "qualified coaching." How does all this work?

We, like our namesake, don't operate in a vacuum. All of our programs are recognized by the state and have been accredited unconditionally by rating agencies and state university commissions. We submitted all of our basic concepts to them and had them approved. We were one of the first German institutes of higher learning to set up our entire program according to the criteria of the Bologna Process, that is to say based on the bachelor's and master's degree system.

But how can I give shape to these vague-sounding demands? I'd like to be able to picture this more concretely.

Take our "tandem coaching" concept. We have each student select two coaches, completely on his/her own, an academic and a practical-experience

---

Photo   (**see page 285**) Götz-Wolf Wagener, President of Columbus Holding AG

coach, who will stay with the student for the duration of his/her studies. It is a required part of the curriculum and subject to normal examination regulations. At least once per semester the student is required to meet with each of these coaches and work through an agenda: how am I doing, am I satisfied, how am I proceeding with the goals I've set for myself? And, of course, all this is then committed to paper.

What role does the practical-experience coach have; what role, the academic coach?

The academic coach gives the student ongoing guidance on curriculum-oriented questions, personal development, and professional perspectives. The role of the practical-experience coach is to be a constant reminder that knowledge and skills acquired at the university will some day have to be applied in practice.

What are the minimum responsibilities of the coach – just to sit down with the coachee twice a year, or is there more?

The coaching concept is just the framework, within that there is plenty of space for individual agreements and rules. How the practical-experience coach and the coachee set up their relationship in detail, how often they meet, where they meet, which topics they discuss, etc., is something the two partners establish together at the beginning of their coaching relationship.

In what way are the students expected to contribute to the process?

The students prepare for coaching sessions by reflecting on the last six months and writing a so-called "semester reflection paper." This focuses on two questions: What was important to me in the last coaching session? What goals do I want to use this coaching session to achieve? The coach then uses this reflection paper to prepare him/herself for the session.

Does the coach–coachee relationship sometimes continue to develop even after graduation, for example do coachees sometimes find a job at the coach's company?

Sometimes these relationships last a lifetime. But they don't have to. What often happens is that a coachee will go on to do an internship at the coach's company. It's still too early to say whether or not students will choose to start their careers from there or start work later.

That means your coaching concept is a kind of open structure: the students are free to choose whichever coach they please, but they have to find him/her on their own, and they have to be accountable to the process.

Yes, the idea is also to get students used to the process of coaching, which we consider extremely important. A coach can help deal with crises and point out challenges and corrections by asking questions. The ultimate aim is to show people how to guide themselves. Our university also profits from this process because it gives us an enormous amount of contacts. The academic coach and the practical-experience coach have to meet once a year to discuss their coachee, and this meeting often takes place here at the university. All of the people involved are extremely interesting people – from mayors to theater managers, gallery owners, and CEOs, to editors-in-chief and opera directors.

Self-organization or: at a university nothing should go unquestioned

You've introduced another open structure that seems even more unusual: the principle that if ten people request a class, no matter what the subject, the university is obligated to teach it. How does that work?

This is based on the idea that we don't believe our programs of study are the ultimate in wisdom; after all, we drew them up ourselves. (laughs) **The students know what they want to learn better than we do.** *We always say: Don't complain, do it yourself! That is the basic principle at our university. Nobody complains here. So if someone wants a class on a specific subject, he has to arrange it himself: define the subject matter, look for someone to teach the class, find nine other people who want to take the class too – and then we finance it. And the subject matter becomes part of the official curriculum. In this way students are invited to coordinate part of the exam-relevant requirements.*

Is that one of your basic principles? If you have an idea, you are free to implement it, but you will have to work for it?

Yes, but we have also implemented a number of other "risky" unconventional ideas. One of them concerns how the university governs itself. Here we have introduced what we call "zone" architecture.

There is a phenomenon in many institutions – and especially in academic ones – where you analyze and discuss problems intensely and then just leave everything the way it is. *With us, if someone wants to change something, he has to set up a "zone," that is, a temporalized work unit in which a resolution proposal is worked out to be submitted to the senate, the university's highest political organ. The senate is required to deal with the* *resolution proposal. You don't have to belong to the university to be able to set up a zone. For example, the janitor can say: We want to change the whole structure of tuition or how it is financed. And he can do that, he just has to do it publicly: set up the zone and have it be activated by the senate. From that point on, others can take part. For example, I could go in and say: But I want to have tuition raised.*

You're free to talk about any subject at all?

Yes, as Jacques Derrida might have said, especially at a university nothing should go unquestioned. But sooner or later you need to find a solution. And this is submitted to the senate as a resolution proposal. And at this point the senate cannot return a negative decision. If it disagrees, it must write a replication in which it states the reasons why the proposed solution will not work and points out possible solutions that would. In the second round, however, a negative decision is possible if furnished with the grounds for rejection. That means it is an enormous effort to complain because you're immediately called upon to act. It's a running gag with us: Hey, I have a problem! – Well then, you'd better set up a zone. – Okay, maybe I don't have a problem after all. (laughs) There are a lot of problems out there that turn out not to be problems at all. And if there is a real problem, then the person who comes out on top with his/her opinion is the one who does the work, not the professors and not the president or whoever else.

We've had the craziest things. For example, teaching zones where people discussed the radicalization of the teaching concept and a self-obligation code for students who want to work harder and do more to prepare for classes, and they make binding arrangements among themselves. Or: we reorganized the entire academic calendar. Instead of the classic semester system, we introduced a new division of the academic year: eight months in a row for study and research and then four months for academic projects or start-ups, long internships, or studying abroad, etc. It was a tremendous effort for the whole staff, especially for people with children. But through the zone we were able to discuss all the relevant issues from the beginning; and from there, windows opened up for resolving the situation. And we did it.

How do you come up with these kinds of things?

Thinking. Discussing. Experimenting. Thinking.

But this extreme radicalness is something unique to Zeppelin University, isn't it?

You won't find it anywhere else. And it does have its dangers because we have to be open to compromise. And each of us has to show his or her strong stance to the university. Each and every one of us.

Elite or pioneering university?

You are an elite university. Most of what you do can't be adopted by other institutions.

We're not really an elite university but rather a pioneering university that excels in experimentation. *But it is true that a lot cannot be done anywhere else. For one, because we don't have the constrictions state universities have, e.g. laws governing public offices. And most of these things only work if you're small. That's why we don't want to get any bigger. We've all agreed to limit ourselves to no more than 800 students.*

Nevertheless, we can try to define what is special about these kinds of innovative structures. Was the format of the zone architecture already set from the beginning or did it evolve over time? Or let me ask this a different way: with these types of structures how do you manage to maintain the balance between pinning things down and leaving things open? Otherwise it doesn't work.

We have to maintain the balance between both extremes – the difference and the unity. But that is just the formal view. Structurally, the question is how do we manage to get decentralized intelligence to lead to a centrally intelligent decision. And that for example means taking approaches like integrating the most ignored body of people at universities: the students. That's my only advice to state-run universities: Talk to your students and alumni! They are the ones who experience the university as a whole. And, of course, it's an evolutionary process too. People make suggestions, try things out, make a decision, make adjustments, and the process starts all over again. For example, we started something that we call Development Day: **once every semester we meet for an entire day and just talk about us;** *400 people focused on just one thing. This can lead to all sorts of probable and improbable results. But at the end of the day we ask ourselves: Okay, so what does this mean? And then we translate it into open structures.*

The university as a cultural impetus for regional development

My second interviewee, Götz-Wolf Wagener, CEO at Columbus, has arrived. He was held up by a business appointment. Wagener has delicate facial features and a neatly clipped beard; in contrast to Jansen's eloquent vivacity, he radiates genteel reservation – a manicured elderly gentleman. Wagener, too, has got stuck at the railroad crossing before. In fact with him the situation was even more extreme because the driver in front of him blew his top: "What is this, Candid Camera?" he stormed about, gesticulating wildly. Jansen grins knowingly: "When the train leaves the Überlingen station, the gate has already gone down here. Tell that to people, and they won't believe you." The university president relates how he invited the director of urban planning in Friedrichshafen to come to Zeppelin to meet with him and that he too got held up and arrived late. Of course that didn't change anything: "To us this is a real problem. Imagine the university is on fire, we leap to our feet, ready to take action, but the gate goes down. So all we can do is jump into the lake!"

Columbus and Zeppelin University collaborate on very diverse projects – from your classic customer relations to different forms of sponsoring to collaboration in the cultural area. Can you briefly describe how this came about?

STEPHAN A. JANSEN/ZEPPELIN UNIVERSITY: Yes, originally it was a customer relationship that has meanwhile expanded to include the most diverse forms of cooperation. It wasn't planned systematically, but the seed for this growth lay in our personalities. It all started with a call for bids for designing our website. A very complex matter, partly because we were creating our visual identity in collaboration with a Documenta artist, Ecke Bonk, whose trademark is non-logo and non-design. That makes things difficult for a design agency. But when I met Mr. Wagener it was immediately clear that we both had an affinity for art. That was extremely important because we couldn't pick someone who didn't understand why we don't want our home page in color, why we don't want flash animation, and so on. If someone doesn't feel that himself, he doesn't understand it at all.

As a company, what was your motivation to accept these conditions?

GÖTZ-WOLF WAGENER/COLUMBUS: I really liked the idea of founding a private university here. I'm the kind of person who can get very excited about new ideas, and I could strongly sense what was being created here and that it had to succeed. Zeppelin University's special mentality, its interdisciplinary approach, also the way everything is infused by culture, these are all things that correspond with our company philosophy. We believe that business activity doesn't take place in a social vacuum. We always see the environment around the company too, the people behind the projects. And at Zeppelin University, we recognized this spirit, this creativity . . . I had the feeling that they were totally living their philosophy, and it would just be great to be a part of it.

Do other companies in the region feel this way too?

Yes, I think so. You could feel it this year at the summer party. I thought to myself: Look at this unbroken stream of VIPs! Respect, respect. Everyone was surprised and pleased at how this little sapling had grown. And Mr. Weishaupt [CEO of Max Weishaupt GmbH in Schwendi near Biberach] owns shares in the holding company too, so he must have faith in its success. Everyone around here remembers Bad Waldsee: from 1991 to 2003 there was a private non-accredited business school here, part of it was later incorporated into Zeppelin University. I didn't like its concept, though, and above all I didn't like its advertising. Later, when I got to know the circle of people who had planned all this, I was fascinated that it was infused with a completely different spirit.

What role does Zeppelin University's interaction with the region play?

STEPHAN A. JANSEN/ZEPPELIN UNIVERSITY: To us interaction with the region is, of course, vitally important. It was clear to us from when we first started working with Columbus on our home page that this would be an intensive project that demanded a lot of personal contact, you couldn't do it over the phone. You needed a face-to-face partner.

*But more broadly, this is a region with a strong tradition in engineering and technology, it was hit hard by the war and largely destroyed, there are a lot of immigrant families here; in addition, there is the beautiful landscape with strong tourism and the famous vineyards and orchards, but its weak point is that there is relatively little in the way of culture. And the city is spoiled in the sense that there are two very successful companies here, ZF Friedrichshafen AG and Zeppelin GmbH, the latter of which belongs to the Zeppelin Foundation, which in turn belongs to the city – incidentally, a unique constellation not to be found anywhere else in the world. This is why the infrastructure here is so well developed, there is an exhibition center and airport, etc. But what was missing was a tertiary school. **If you are a university being founded here, culturalization automatically becomes your responsibility. And even more so if your aim is to transform existing society into a knowledge society.** We realized this early on, at first 95 percent of the population had no connection to us whatsoever: What do we need a business and humanities university for, especially a private one? But the other 5 percent, to which Columbus belonged, came to us in no time. And now the lectures we offer at the Bürger-Universität [citizens' university] are so packed that the city facilities are hardly big enough to handle them anymore.*

Zeppelin University as a cultural impetus for the region?

It wasn't our primary goal – more of a side effect. Whereby in the founding phase, and this is interesting too, the social expectations varied greatly. Some said what the region needs is a young generation of talented experts; others said if we want to maintain current standards, we need to bring in a qualified faculty; the city itself wanted a university; others said: Finally a little culture around here, maybe the students will help revive the city center. So the idea of founding a university meant something different to everyone. But hardly anyone actually thought it would be successful. (grins) I think that's changed a little.

How is Zeppelin University specifically networked with the region culturally?

We take an active part in cultural life in many different ways. Our lectures and performances are open to the public, anyone can come. And we are involved in all sorts of things, e.g. through our students' projects. The exhibition "consum," which our students curated in October/November 2004 at the Kunstverein Friedrichshafen in conjunction with international artists, was the biggest-drawing event in the history of the Kunstverein. For several weeks there was something going on all the time. Installations, readings, films; Professor Birger Priddat, our chair for political economy, gave a two-hour lecture on the banana, followed by an original Tupperware party, and much, much more.

Did the idea for the exhibition come from Zeppelin University or the Kunstverein?

The idea for the exhibition came out of a class at Zeppelin University. It started out with the students exploring the phenomenon of consumerism, then they researched which artists would fit in with this topic and invited them. Some of the artists created works especially for the exhibition. And it wasn't long before the Kunstverein was called in to help with conception and planning. The Kunstverein provided the exhibition spaces and, in effect, also its audience – and backed the project with assistance and advice.

You have also developed an unusual method for approving student applications – you integrate the region here too.

You mean our Pioneers Wanted! day. As you know, we only accept 30 new students per semester. So what we do is we send people in small groups to various charities and public offices or to the Columbus Art Foundation, where they are given a real-life task – which usually means an irresolvable problem. A complex task that the institution itself has been working on for ages. We don't have anything to do with the particulars. But the applicants are assessed according to how they handle the situation. And of course this whole event helps to raise public awareness of Zeppelin.

Can you give an example of one of these tasks?

On our last Pioneers Wanted! day in 2006 a few of the groups were sent to the Ravensburg Correctional Facility (JVA). There they were given the question: How can we encourage inmates to take interest in basic everyday practical tasks? In concrete terms they were asked to draw up a social competence training program for inmates as a way of preparing them for life outside the walls. That day we had 30 applicants working on this case study in groups of five. Their first reaction was one of surprise because when they got on the bus that morning, they had absolutely no idea where we were taking them. Then, when we stopped outside the prison gates on the outskirts of Ravensburg, open curiosity mixed with eager surprise. From there on in, it's up to the applicants what they make out of the situation.

What do they do with the ideas afterwards? Is there an example of an institution being able to use the ideas in any concrete way?

Generally speaking, very little of what the applicants present at the end of the case study is really new to the institution. But even this can be helpful: the applicants reinforce an already existing idea.

For example, two years ago we sent a few applicants to the Waldorf school in Überlingen. The school wanted to convert its tuition system from an individual model that took the income of the parents into account to a system with one set tuition fee per child. The task was how to communicate this change to both the parents of the current students and the parents of prospective students. And in this case we know that the teacher in charge of implementing the tuition change did use the instructions and recommendations given by the applicants. Sometimes institutions call up months later to report that in connection with a new project they remembered one of the ideas and were able to put it to good use. Meanwhile word of this method has got around in the region, and sometimes institutions approach us to ask if we could include them too.

Practicing business

Have cases like these sometimes changed an institute's relationship to Zeppelin University?

Yes, Columbus is a good example of this. To date we have held three Pioneers Wanted! days there, and since that time the level of cooperation has changed, become more intensive. And some of the students who are currently curating the project "anders sehen" [seeing differently] have known Columbus since the time they spent there on their Pioneers Wanted! day.

The project "anders sehen" is a good example of very close collaboration between a university and a company. Is this sort of thing only possible if there is a special personal relationship like the one you share?

GÖTZ-WOLF WAGENER/COLUMBUS: Yes, I would say so. It's an idea that developed between us at Columbus and Jörg van den Berg, who was still with the Zeppelin University Arts Program at the time. And the way it worked was that our curators chose roughly 50 objects from the Columbus Art Foundation inventory, and from this pool, three groups of students were given the task of each putting together their own exhibition. Complete with concept, installations, catalogue, everything.

And none of the groups were allowed to know what the other groups were doing. It was an assignment that counted toward the final exam. Out of this emerged three exhibitions, which ran successively from October to December 2006 at the Columbus Art Foundation. The first one was entitled "co.existenz – natürlich künstlich" [co. existence – naturally artificial], the second "oberflächlich – aus Tiefe" [superficial – out of depth], and the third "alles inszeniert" [everything staged]. And the exciting thing was seeing how fascinated everyone was as the students put the exhibition together, even the company employees. They would come over and check on progress. Suddenly, artwork that normally just stood around in storage was given new life. And sometimes the situation pushed the envelopes of both sides involved because you couldn't do things the way you normally would. For example, maybe they needed more money for this or that and we had no set procedure for dealing with that kind of thing. Or in some cases the students had to acknowledge that some things just weren't possible. It was extremely stimulating for everyone. And that's something that developed out of the specific opportunities provided by our collaboration.

Your company earns its money by providing office, IT, and media services. What role does your commitment to art play in the company's image of itself?

A significant one. It has become part of our profile – in our sector we are referred to as "the art guys." My commitment to art is personally motivated: my daughter is an artist, I saw how hard it is to get started after graduation. And that's why we developed a program to foster young artists. And this even earned us the German Prize for Cultural Sponsoring. We support young artists starting out in their careers, usually just after graduation from the art academy. The way it works is we select two to three artists and follow their progress for two to five years. We offer them a kind of package: the Art Foundation acquires a portfolio of the artist's work, finances a high-quality publication for presentation purposes, and helps with contacts and organizing exhibitions, etc.

In this way we have fostered some 35 artists to date, and we even continue to provide advice and support to some of them.

The curator Jörg van den Berg, who you mentioned earlier and who used to run the Zeppelin University Arts Program, recently "switched sides" and is now running the Columbus Art Foundation. Is that the superlative example of close collaboration between a university and a company?

It is definitely the sign of a good relationship. I know that I'm getting a good man.

STEPHAN A. JANSEN/ZEPPELIN UNIVERSITY: *Yeah well, you do what you can to help. (laughs) But seriously, there were a number of reasons for the switch. In a way it's normal for us to want to draw people from outside the region and later release them into the region. We regard that as part of our job. It's the same with our students, and with our staff too. But we only voluntarily give up good staff to good partners.*

Thank you for the interview.

Key project insight

- The collaboration between Zeppelin University and the company Columbus is diverse, from classic customer relations to different forms of sponsoring to concrete cooperation on art projects with the Columbus Art Foundation. Columbus provides financial resources, manpower in the form of curators or staff, or makes artwork temporarily available. This close cooperation is based not only on a strong shared affinity for art, but also on the personal trust that has grown between them over the years, and a commitment to the region.
- Zeppelin University itself is extremely innovative and enjoys experimenting in everything that has to do with drawing up the curriculum, the responsibilities of the students and professors, and university organization. Our highest principle: if you want to improve something, you have to take responsibility yourself. Complaining gets you nowhere. The goal is to turn the university into a "learning organization."
- While much of what Zeppelin University does can't be achieved by other institutes of higher learning for structural reasons (small size, not bound by the laws governing public offices), certain aspects, if adapted accordingly, could be adopted by some institutions – for example, practical-experience coaching or the annual Development

Day, an institutionalized setting that nurtures ideas for the continued development of the school. Another exemplary aspect is the wide range of information offered to prospective students, for example Rolling Admission, Incoming Procedure (on the ZU website), or the annual information day where visitors can find out about every aspect of the university.

Who or What is Innovative?: Basic Scientific Research on Cooperation

Interview by *Hans-Joachim Gögl*

Switzerland

"I spend a lot of time with stimulating people. But an inspiring encounter doesn't begin with specific intentions; rather, the conversation's own dynamics are what make it fascinating. In that sense I wonder to what extent it is even possible to instrumentalize collaboration."

A basic research project in business management in cooperation with companies from the sectors of software engineering/information technology and life sciences/pharmaceuticals.

RISE, a research center at the renowned Swiss University of St. Gallen (HSG), deals systematically with getting to the bottom of the capacity for innovation: how should a company, an organization, be set up if it is to be capable of constantly developing and successfully implementing new insights? A key topic for company and university researchers and developers.

Beyond that, this cooperation model might also offer an unconventional exemplary approach for other disciplines by showing how basic research projects in cultural and/or social science areas can under certain circumstances directly affect companies and also be funded by them. And that, although RISE doesn't offer any practical solutions to problems and insists on years of preliminary cooperation.

An interview about cultures of innovation, cooperation, and applied basic research.

Dr. Simon Grand, founder and academic director of RISE Management Research at the University of St. Gallen. Parallel to this, he is also active as an entrepreneur – most recently as partner and chairman of the board for an international design and brand agency – and advises and assists businesspeople, executive boards, and management bodies in their strategy, business, identity, and innovation processes.

Stefan Arn (Dipl. Informatik-Ing. ETH), founder and CEO of AdNovum Informatik AG and president of ICT Switzerland, the umbrella organization of the Swiss information and communication technologies sector. In 2003 Stefan Arn was named Entrepreneur of the Year by Ernst & Young. He has run the Stream Delivery Unit Clients & Products at UBS AG since the beginning of 2007. At the time of this interview, Stefan Arn and his company had collaborated with RISE and been the subject of its research for roughly five years.

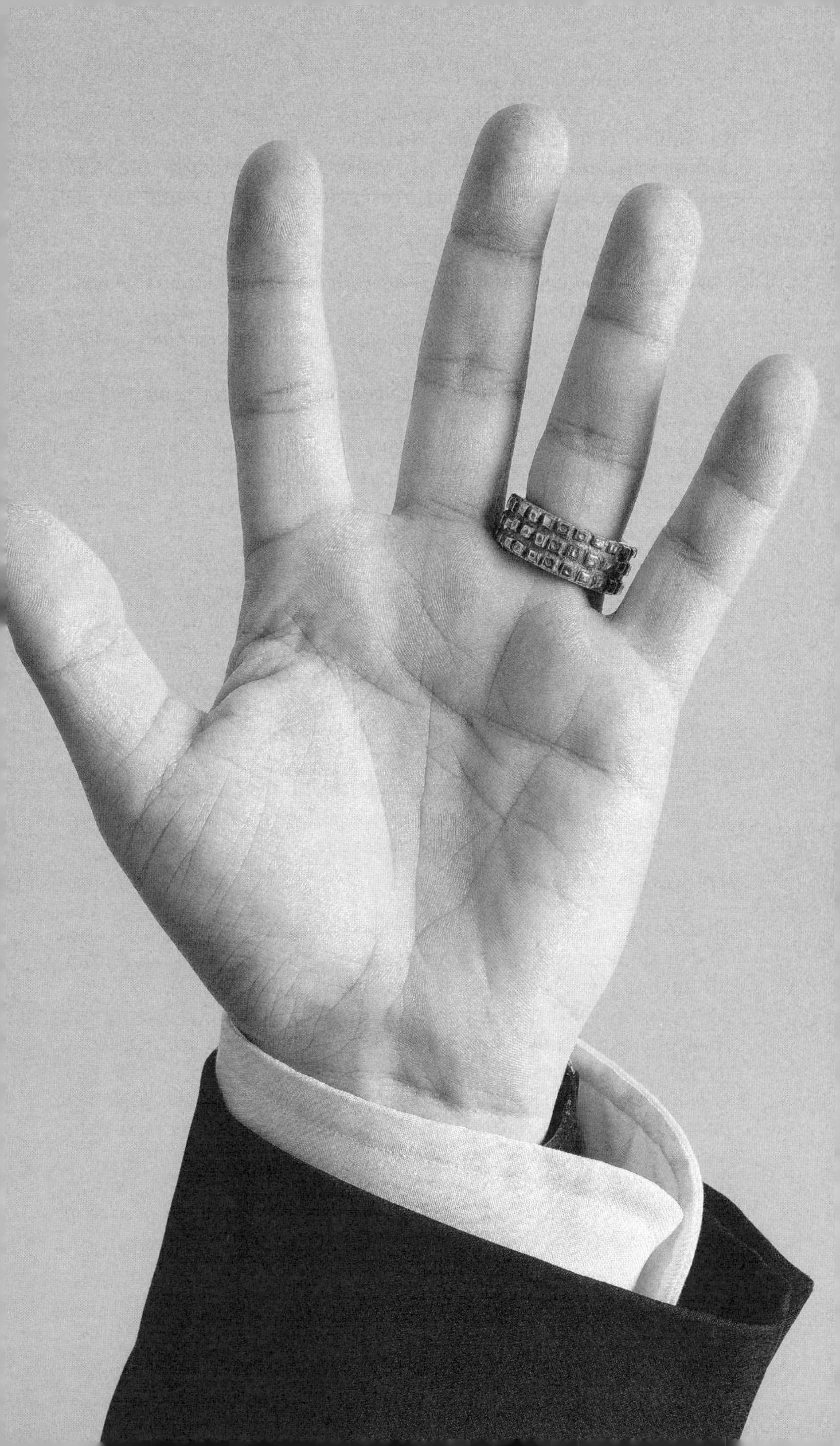

Dr. Grand, as researchers you systematically study the conditions that allow companies to be innovative, to bring new things into the world. In the process have you come across executives with extraordinarily high cooperation capabilities?

SIMON GRAND/RISE: Yes and no, and that's because cooperation has to do with coming closer and innovation thrives largely on the crazy, the unconventional, the not immediately understandable. These factors aren't necessarily conducive of cooperation, but for the ability to create something new they are absolutely fundamental.

STEFAN ARN/ADNOVUM: Long-term cooperation projects thrive on the strong curiosity the partners have for each other. In the 19 years of experience I have had working on this subject it has always been the same: the preconceived notions the partners have when they start working together always fade over the course of time. If new topics don't move in to fill the void, the partnership always dies.

I am sure there is no axiomatic connection between cooperation and innovation. Many excellent researchers in the chemical industry, for example, are incapable of cooperation because they are fully focused on their topics and are not interested in interaction. **The will and ability to work together is widespread among generalists who are convinced that new solutions lie beyond their scope of vision.** *To them surprise is the most stimulating thing in life, and this gives rise to curiosity, which drives collaboration.*

Thus, with many research companies it is not actually the researchers and innovators themselves who are behind cooperation efforts; instead, there is a third force at work that translates and connects.

That means innovative companies make institutionalized "spaces" available to allow knowledge and insight from different disciplines to come together and reach, in a manner of speaking, a consensus?

SIMON GRAND/RISE: The thing about all innovation processes is that you never know what is going to come out in the end. And even after you've looked at the first results, you can't tell if a newly developed product is really going to prove successful. One of the functions of cooperation in connection with innovation is dealing with this uncertainty. I have a name for these spaces you mention, in reference to a science historian I call them "trading zones." Actually, they are bazaar situations where you can trade notions, concepts, ideas, and where people get together to confirm, reject, deduce, and also decide what is innovative.

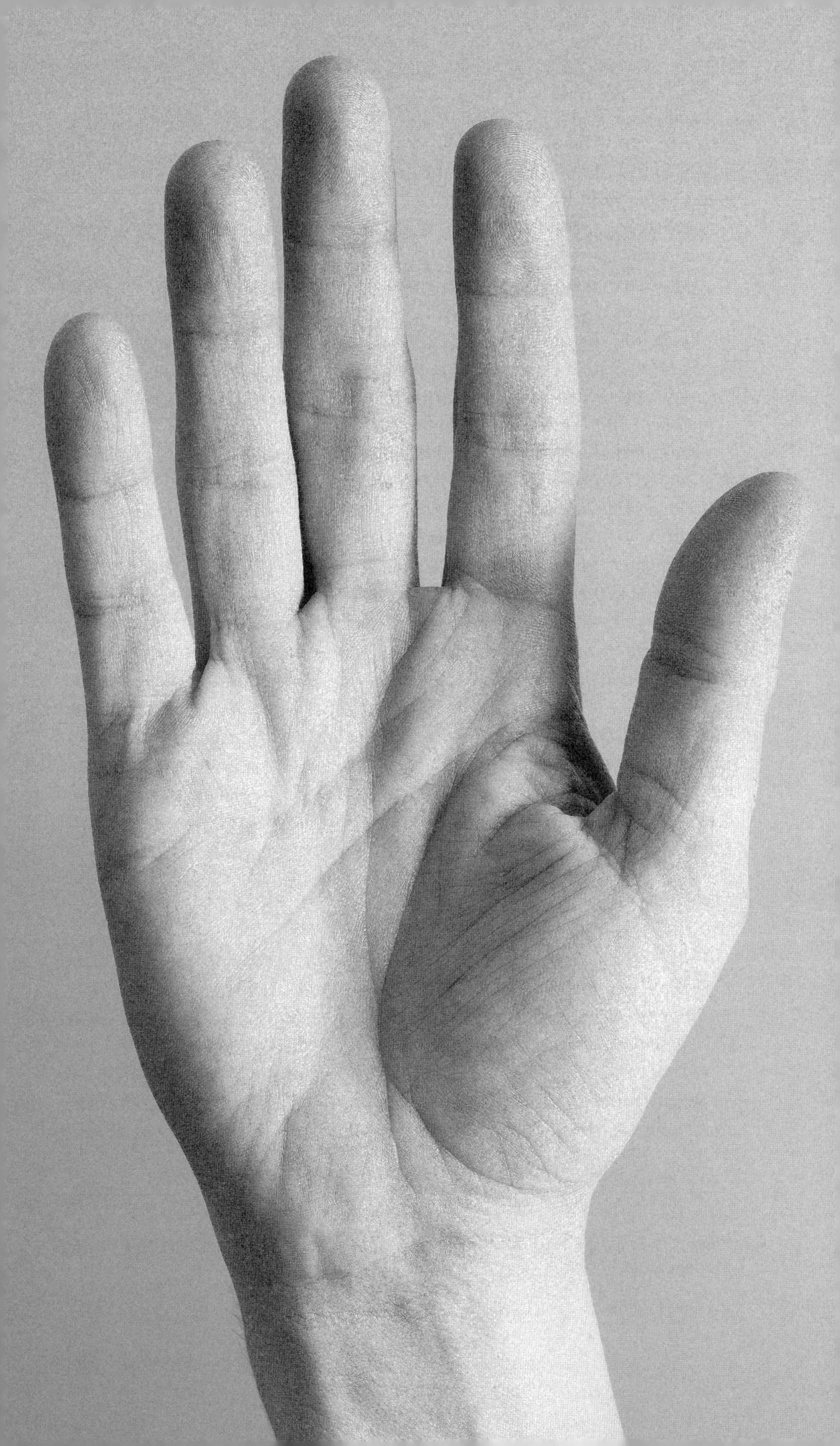

Mr. Arn, isn't the ability to cooperate and interest in doing so biographically founded and fed by what one has experienced, lived through, and been personally influenced by?

STEFAN ARN/ADNOVUM: I never liked teamwork as a child. On the contrary, I was always a strong individualist, but I don't think that's really the point. Cooperation in connection with technological innovation is an intellectual matter. In America, group work, interdisciplinary cooperation, and cross-fertilizing are built into the university system. I attended classes for two years at Stanford and was neither positively nor negatively impressed by these methods – though I collaborated with two attractive biologists on a social level, I cannot remember what their scientific contribution was.

I only started to grasp the value of cooperation when I realized how difficult it is to find someone not involved with my topic who was able to introduce any halfway usable new ideas. These kinds of benefit-oriented relationships always tend to be very conscious, temporally defined decisions that don't come from an immanent fundamental conviction.

SIMON GRAND/RISE: A follow-up thought to that which by no means contradicts your pragmatic focus is: I spend a lot of time with stimulating people. But an inspiring encounter doesn't begin with specific intentions; rather, the conversation's own dynamics are what make it fascinating. In that sense I wonder to what extent it is even possible to instrumentalize collaboration. And wouldn't I have to constantly expect things to come out differently, expect topics to take a different turn than I originally assumed? Because that is when something new, surprising, unexpected happens. I may have my expectations going into this encounter, but we actually build whatever it is we're dealing with together in the situation itself and from there we follow the dynamics of our relationship.

Without this attitude, this type of encounter would be an exchange of information instead of a creative situation.

Within the framework of RISE we only collaborate with companies with this open attitude. But we have also often observed that if someone has a lot of experience with innovation, he knows that coincidences and unexpected developments are crucial to coming up with a really good new idea.

STEFAN ARN/ADNOVUM: That is definitely a prerequisite. Many researchers are so focused on their topic that they are not touched at all by anything else. They don't even feel like anything is missing. And if you don't perceive an absence,

---

Photo   (**see page 309**) Simon Grand, founder and Academic Director of RISE Management Research

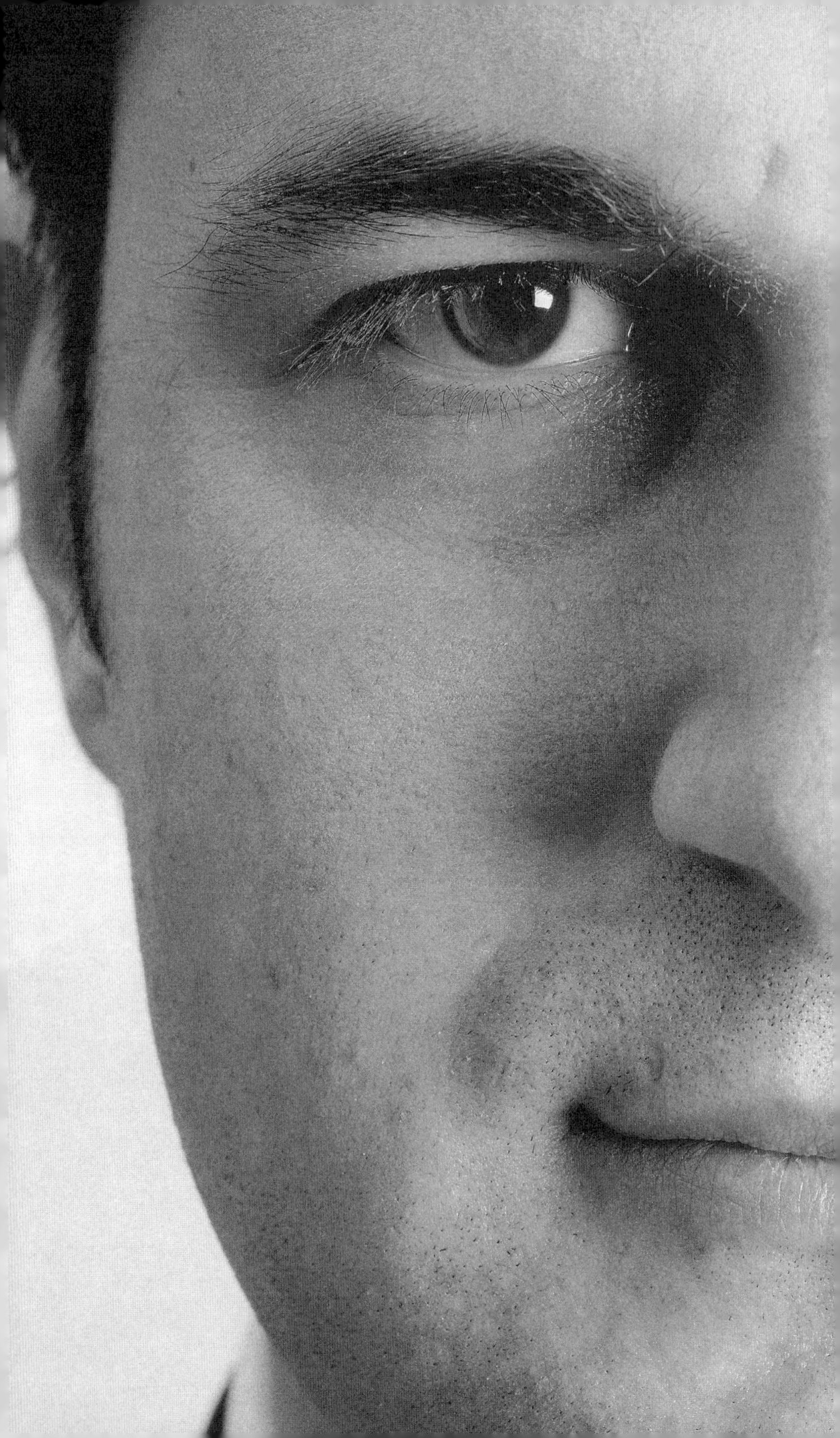

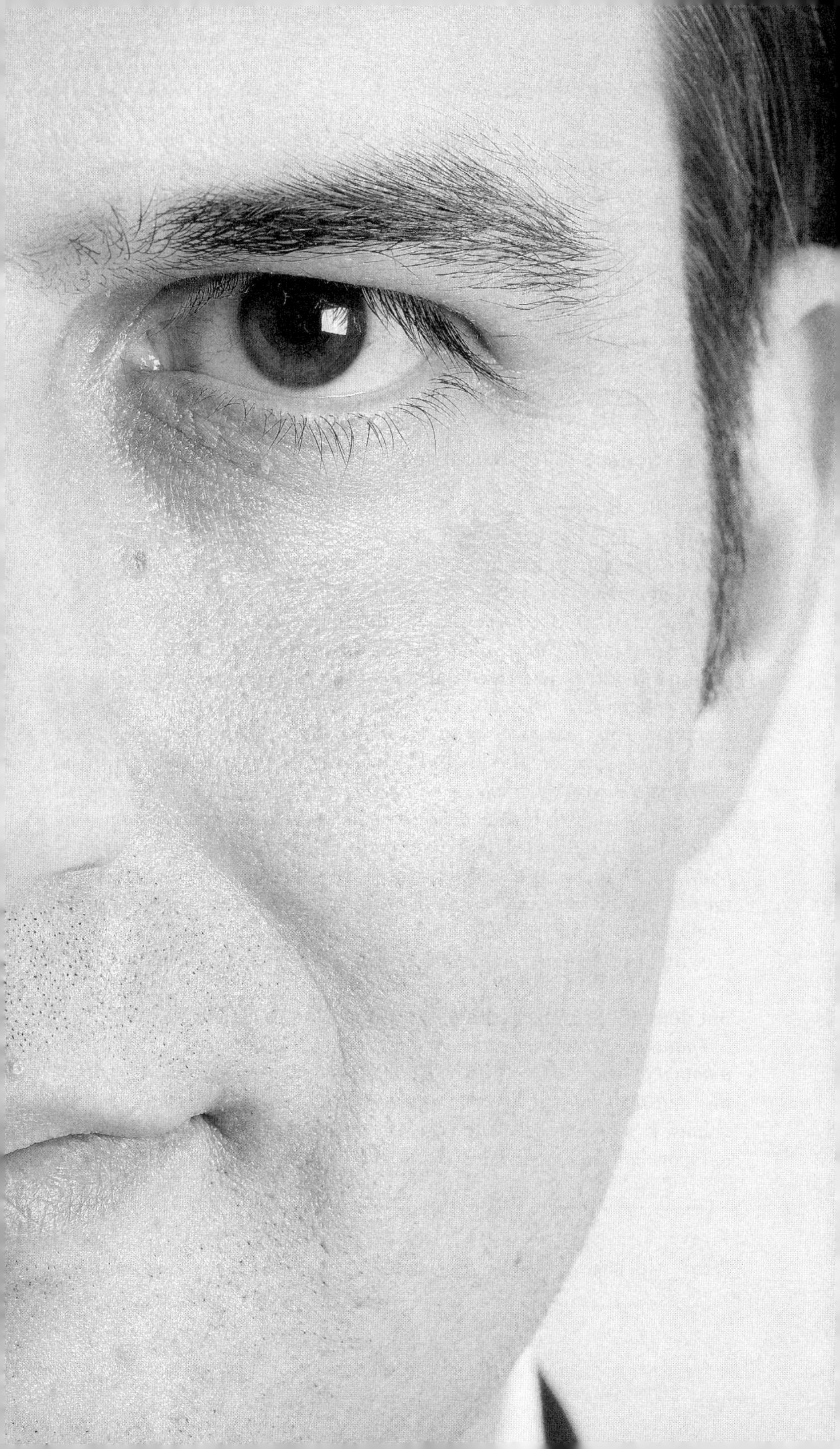

there's no solution. So, a priori, they are less than amused when someone like Simon Grand starts telling them about French philosophers when as entrepreneurs all they want to hear about is growth. You've got to be curious by nature about unconventional approaches to a subject and believe in the possibility that an approach like this has the potential to solve your specific problem. And it does! Modern research on semiconductor technology and biochemical processes was only made possible through lateral thinking. If we had just continued to delve deeper into our own highly specialized knowledge – the logical and accepted fundamental beliefs of experts – no one would have ever come up with the idea of storing information in crystalline structures.

Innovation: defining the concept together

Now, this attitude or sensibility can be promoted or impeded in organizations. To what extent is it important for research-oriented organizations to reach an internal consensus on what constitutes innovation or progress in this culture?

SIMON GRAND/RISE: To begin with, our studies revealed that each company has a different concept of innovation. **But for organizational cultures a shared notion of what innovation means is absolutely essential** *because a definition of innovation can never be free from bias. One finds certain things good and others bad, that needs to be set straight. Of course there are gray zones where people disagree. But essentially we agree on a core value system and that makes it possible to have a common understanding of what is perceived here as progress. This is generally defined by the pioneers of the company, who establish an initial nucleus that continues to influence the culture almost automatically for years. As a company grows, it hires new employees and despite the fact that they were not part of the discussion about values, they still function naturally within this framework. Whether or not something is innovative is always relative to the given value system. There is no absolute definition.*

But doesn't the market ultimately decide what is innovative?

That's an interesting question. What I've observed is that companies develop a notion of what is innovative based on their own views and values, and then they test it to see if it is viable in the market. Market feedback is essential because it allows the company to explain and better justify its own conception of innovation to the customer.

Photo   (**see page 313**) Stefan Arn, founder and CEO of AdNovum Informatik AG

STEFAN ARN/ADNOVUM: Executives have to discuss a coordinate system within which the actors are to operate, but they also have to demonstrate within the company what progress is. For example, in software development there may be 20 different ways of implementing a task, and the market will buy ten of them. But for the internal dynamics of the organization you need this set of values, otherwise nothing happens because different forces are all pulling in different directions.

SIMON GRAND/RISE: To illustrate this with an example: there are cultures that have an evolutionary understanding of innovation. Innovation is always relatively close to what I already know, and I only take risks at very specific points. The creative process in this organization has a completely different structure from that of a group to which innovation means radically challenging existing industrial standards. Within a given community you will find both of these approaches and views, and they have enormous organizational consequences because they also help define who in the group is cool and who is uncool. And both approaches are legitimate; it has nothing to do with which of the two cultures is more economically successful.

That means that one of the main tasks of a manager in a research-oriented organization would be to establish a collective concept of innovation?

Exactly, what all innovative organizations have in common is an official, pronounced, and well-propagated understanding of innovation.

STEFAN ARN/ADNOVUM: And that is an ongoing communication process that starts at the initial interview and is a subject in almost all subsequent meetings. We have a very clear set of priorities as to what constitutes innovation and what doesn't, a list of values that is constantly being adapted, but which everyone is familiar with in detail. To us, for example, innovative means when someone keeps deadlines, sticks within the budget, makes use of existing recyclable elements, and creates new ones that can be recycled in the future. Developing an ingenious yet still unstable crypto algorithm might attract attention within the company, but we do not see it as innovative, and everyone at AdNovum knows that.

The way the two of you talk, the charm of invention is reduced to an assiduous administrative chore. But the question still remains, how do you arrange for the kiss of the muse, for inspiration?

*That's not the tough part! I can always find something new. Really good researchers have brilliant ideas every day. **The tough part is bringing the good ideas down to earth.** Go over to our coffee bar, you'll be dished up ten new ideas within 15 minutes. But try to find someone who can bring one of those balloons down to earth and get it to work within the rigid framework* *of a project. That's the real achievement and the expensive part of the exercise!*

I've had great ideas, especially while sitting in boring meetings, and you just have to make sure you don't forget them. The decisive task of the manager, however, is to create a process for transforming this plethora of ideas into one useful market-viable solution.

SIMON GRAND/RISE: That is one of the essences of innovation research, and historically it goes back a long way. Joseph Schumpeter defines innovation precisely so: innovation is invention plus execution. That means having good ideas – he even assumes that statistically they are evenly distributed around the world and over time – and then actually executing them too. Whereby, the main challenge is market implementation or execution in a sociopolitical context. And for this a specific communication capability is important because I have to be able to talk about something new and unconventional in a way that its logic and usefulness is credible to other people on a practical level. This aspect is vital to every company and a key element in competition because you will always have rivals with a different standpoint and good reasons to back it. **Talking about innovation is a fundamental part of its execution.**

That means that from this point of view innovation is first and foremost a power struggle of who can prevail with his/her interpretation of what constitutes new or better without there being any objective definition at all: it's all about who can communicate his perspective more convincingly within his own organization and to the customer.

STEFAN ARN/ADNOVUM: Exactly! A wonderful example from the software industry is that the worst possible operating system asserted itself successfully on computer desktops: Windows is an aggregation of mistakes that runs more or less by chance. The counter-example is Mac OS, which despite being stable has failed to assert itself. That means better or worse is not necessarily a market criterion.

SIMON GRAND/RISE: The companies in our study that rank among the best in their sector have all set down a strong system of values on which they orient their actions. Only if I have points of reference to guide me can I make decisions or even take action in the first place. Because if it is true that innovation is inherently uncertain, I have a constant orientation problem. What am I supposed to orient myself on, what is right and wrong? The companies in our study, for example, know exactly what success means to them. You have to have a clear idea of this because if to me success is doubling my sales volume every year, the company will develop in a completely different direction than if success means a) every project reaches completion, and b) I grow if I manage to generate further interesting projects. Another important system of reference is quality. What demands do we make of ourselves in the things we do? This value system for innovation, success, and quality acts as an identification gauge both outwardly and inwardly.

And because one can just as easily hold a different view, I as executive manager have to stabilize these reference points. How do I do this? **Empirical research shows that entrepreneurs spend more than 95 percent of their time talking.** *In this way they strengthen these beacons of orientation and apply them repeatedly in new situations and problems.*

Communication, competence, and academic training

Mr. Arn, based on this, how do you assess the university training of engineers who come to work for you after graduation?

STEFAN ARN/ADNOVUM: In Switzerland there is an association called "engineers shape our future" (www.ingch.ch). Its members are renowned companies such as Swiss Re, ABB, Hilti, Siemens, and AdNovum, among others, who all have a vested interest in engineers getting a good education. Last year on the occasion of its 100th anniversary we presented the Swiss Federal Institute of Technology Zürich (ETH) with a 25-page documentation in which we described what we expect the education and training of modern engineers to entail. We must have stated 30 times that today's engineer doesn't receive enough cross-disciplinary and communication training. That's where the absolute greatest deficit lies and that's where we need to invest most!

SIMON GRAND/RISE: The subject of communicative skills is an interesting issue that is sometimes discussed in a strangely theoretical way, but what exactly is meant by it? I can only describe which aspects I notice when I talk with entrepreneurs and managers. One of them is that experienced executives know how to listen. That's something, for example, that I would develop! A second aspect is that they are good observers: that means they can, in a relatively precise way, see things and describe them, and they do so without necessarily passing judgment. They know how to ask questions.

Sooner or later, in every interview I conducted with an experienced entrepreneur he started asking me questions. They want to understand the underlying circumstances in detail. Incidentally, I think that is one of the reasons they spend so much time with communication and interaction, because when they talk to other people, they don't just deal with subject matter but also with the underlying reasons, experiences, implicit assumptions.

They can give reasons, explain, and justify rather than just claim why they have a certain opinion and not a different one. Excellent executives are facilitators, that is to say they can guide others constructively in their process, so that they can advance as a community.

I could go on, but what this study demonstrates is that we have to take communication competence apart in order to understand exactly what it means, and that's when I find something really interesting, something that surprises me.

I think a good academic has to be able to do all this too, and these skills could also be taught as part of the university curriculum, regardless of whether the student went on to pursue an academic career, work for a company as an engineer, or become an entrepreneur.

Where are the experts for listening, asking focused questions, and giving convincing reasons?

You used to be able to find them among the philosophers and theologians. But these fields have meanwhile become so scientific that even they have lost these competencies and skills. Oxford and Cambridge have retained relics of this culture. And funnily enough, it's more common in the English-speaking than in the German-speaking world. Of course, through immigration, American campus culture was strongly influenced by the European university tradition. But it's a strange twist that for a number of reasons we Europeans reproach Americans for lacking this very tradition – which in the case of oral culture and scientific debate simply isn't true.

In this context what would you as an entrepreneur wish for from the universities?

STEFAN ARN/ADNOVUM: Once over dinner, we talked about a training course for entrepreneurs and came up with exactly the same points: how do I observe, how do I ask questions, how do I argue a standpoint? This is crucial in the software industry because our discipline is just a kind of toolbox that we adapt to a given field to solve specific problems. We're constantly working in areas that are not our field of expertise – my company, for example, in the field of financial service provision.

There is this traditional notion that a software engineer should have a broad knowledge base and at the same time have a special field of expertise. This is the classic general studies concept, and it should be kept as it is. In addition to this, however, we now need to place greater emphasis on electives in which students are required to apply this specialized knowledge in a specified way so that they will understand the importance their major plays in different fields. And also – and as mentioned above, this is something American universities do quite well – we need to practice presenting the results of our work to a large group from our freshman year on. In America, students have to present their homework to the entire class from week one on.

Most computer science majors also have a minor, usually in business administration. I find it more advantageous to choose a minor that allows greater lateral thinking. We have employees who took medical engineering, another employee minored in geotechnical engineering, someone else had agricultural construction, my minor was theoretical linguistics. To me this is much more fascinating. The more unrelated the better.

In our line of work it is important to be able to quickly grasp foreign terminology, observe and anticipate customer needs, build a prototype, and convince the customer that this is the solution to his problem.

What distinguishes the best engineers is that they are very good at precisely these things. Of course none of this does any good if you don't have a good technical background, and this can be very demanding and often gets neglected for soft skills. We need both.

SIMON GRAND/RISE: To this I'd just like to add that ETH Zürich's great philosopher of science Paul K. Feyerabend always used to say that one shouldn't teach just normality but also dissident knowledge. In every branch of science there were certain notions that didn't manage to assert themselves and others that did, and the ones that did are now a standard part of the curriculum. What he proposed was that we also discussed those subjects that drew the short straw because that way I learn something about how we draw the borders between what is valid and what isn't. That would lead to a creative productive way of dealing with one's own conventions, and the existing borders would be constantly subject to renegotiation. An idea that makes a lot of sense!

Basic research on actual practice

Mr. Arn, you have been collaborating with RISE for the past five years within the framework of this business administration basic research program. I find this extraordinary since RISE doesn't offer its research subjects or partner companies any consulting services or applied strategies for solving their problems. As an entrepreneur do you see your collaboration as a form of idealistically sponsoring this academic institution or is it perhaps an educational project?

STEFAN ARN/ADNOVUM: An educational project, but also a form of consulting input, one that doesn't follow the usual consulting model. The traditional consultant will finish up by showing you, typically on nine transparencies: This is the initial situation blahblahblah, and these are the five solutions. That is exactly what RISE doesn't do, they say: We have observed this at your company and the other partner companies, and this is our explanation, these are the facts, these are our findings. We then discuss all this with the other companies at workshops. To me it was intriguing to discuss the comparisons between our company and similarly structured companies with the other executives.

What I do with all this information is up to me. The researchers remain neutral. I am given a vehicle for thought and in addition to the up-to-date insight I also receive selected, relevant theoretical knowledge that lets me continue to develop – and that is tremendously valuable. And of course this collaboration has a very concrete effect on the company itself. In principle, understanding what is behind the findings is more important to me than the findings themselves. After all, there is nothing more idiotic than an accounting course for entrepreneurs. An entrepreneur should use his time to earn money and hire an accountant. And if you go to the Maldives for a 7,000-dollar, three-day seminar (including a sailing excursion) to learn about the latest entrepreneur fad, you might hear that some practitioner pundit has a new scheme for motivating people. But you still don't have any idea what that scheme is. They never go into detail about that kind of thing in the usual management literature because there isn't any theoretical social science background to draw from.

SIMON GRAND/RISE: Jim March, one of the most important scholars on organizations today, recently said in an interview, and I paraphrase: It's not up to me to decide whether a scientific thought is relevant in practice or not. That decision is up to the person who has to solve a task in a particular situation and within his specific field. And more and more often business executives with plenty of experience are telling me: I'm not going to let someone else tell me how to make business decisions. This is exactly the extremely questionable premise underlying so many consulting situations: someone else is telling me what to do when he can't possibly understand the complex background of either the executive or the company, and without assuming any responsibility himself.

We describe or evaluate only in relation to the existing scientific literature and offer an intensive process for reflecting on the insight gleaned, but we don't give any suggestions for taking practical measures.

STEFAN ARN/ADNOVUM: And that is precisely the key advantage and the main reason why we want to work with RISE.

Mr. Arn, after so many years of experience what would be your advice to an entrepreneur who was thinking about starting a research partnership with a university?

First, as we have already mentioned, to not start the process with a too focused and closed mind. To be willing to go with the flow, to be open and alert to unexpected insight. Second, you have to send in your best people. It's not enough to say, okay we're working with the ETH, with institute so-and-so, and we have this developer sitting around with nothing to do, let's get him to coordinate the project. Wrong! **You have to send in extremely good people and**

a high proportion of them. Don't send the junior but the senior partners. It might be expensive, but my experience has been that this is what successful cooperation projects with universities is based on!

Dr. Grand, doing basic research in cooperation with companies and on top of that being co-financed by them is a stroke of luck that a lot of university researchers dream of. From a structural point of view, is RISE a model that could be adapted to other disciplines?

SIMON GRAND/RISE: Definitely, I encourage it and welcome the chance to learn from other people's approaches. But I would say there are a few basic requirements for this type of model to work. On the industry side, for example, you need an already existing understanding for academic–technological solutions. That means, as Stefan Arn put it, **making your contact people top executives who do not consider examining the premises of their own thinking an obstacle but a prerequisite.** *And then, of course, you need a long-term cooperation agreement with the participating companies.*

Another essential element on the partner side is that rival companies are willing to submit to comparative case studies and to listen to and discuss the results together. In the field of basic research this is more likely to happen than in applied research. But the executives involved have also realized that if they aren't willing to deal with their rivals, they can't continue to develop successfully as entrepreneurs. This attitude is essential for partners who want to work with us.

The PhD students conducting research at RISE all have practical experience, which is very, very important for cooperation projects. In addition they also have an affinity for technological subjects. Finding researchers at the university who fit this mold is, I think, the most difficult part of this model.

What do you consider the scientific challenges of this model?

One has to be aware that the partners on the company side of this cooperation project have a paradigmatic role. Within the framework of our study what emerges as entrepreneurship has a lot to do with Stefan Arn and this company as well as with the other partner companies.

Is that a strength or a weakness?

With the kind of research I do, I don't think it is possible any other way because I have to build a model and observe and understand it precisely and in-depth over a long period of time. After that we can add other cases, in which many things might be solved differently, and we start to recognize and explain these differences. And this, of course, will affect the first company and pose new problems and questions. But through all this, there is still just the one point of departure from which the research agenda develops.

STEFAN ARN/ADNOVUM: Another way of doing it would be to depart from the relevant literature, to use this as a system of coordinates, and then look at ten companies and fit them back into the structure. But the exciting part is to depart from a real model instead of confirming existing scientific premises!

Thank you for the interview.

RISE Management Research

A research center at the Institute of Management (IFB) at the renowned Swiss University of St. Gallen (HSG) which conducts research on the importance of entrepreneurial practice for the commercialization of research and technology in cooperation with companies from sectors in which research & development – an ongoing transformation process from inventions to new products – is existential. Some six researchers are involved in this project, which deals with the systematic investigation of innovation capacity. What motivates these organizations – large corporations such as Roche, but also SMEs and spin-offs – is the constant involvement in a scientifically founded reflective process of their organization, of their innovation dynamics.

AdNovum Informatik AG

A leading Swiss software company specializing in high-end security, application, and integration projects. It has a workforce of roughly 140 employees. The company headquarters is in Zürich, and it also has offices in Bern and Budapest. Its customers include leading Swiss banks, the Swiss Post, federal government agencies, and regional administrative bodies.

Key project insight

- An ongoing process of establishing the values of a common understanding of what constitutes innovation within the group, whether it be a university or commercial organization.
- Cultural and social scientific basic research is relevant to commercial businesses and helpful to them in practice.
- Basic research on collaboration projects as a self-empowering approach to management consulting.

- The creation of useful and exclusive platforms for communicating scientific results to the cooperation partners: workshops – sometimes even with rival companies – in which executives and scientists reflect on the research results.
- Consistent definition of the rules of collaboration: what we as researchers offer and what we don't offer, working together in the long term, financing expectations, the company's contact people the time investment to be expected.
- Send your senior partners: cooperation projects with universities should be conducted with high-level executives from the company side.

Contributors:
Self-portraits

Claudio Alessandri, Photographer

Born in Ferrara, Italy in 1955. At 16, sold his birthday present – a motorcycle – cheap so he could buy his first camera. Claudio studied architecture at the Milan Polytechnic, where from 1975 on, he assisted such trendsetting and now legendary photographers as Roberto Carra (art director of the Italian *Vogue*), Barry Lategan, Oliviero Toscani, and Norman Parkinson. In Cinecittà he witnessed the shooting of Fellini's *La Città delle Donne* and Ettore Scola's *Il Mondo Nuovo* as an illegal observer. In the early 1980s he became Gian Paolo Barbieri's first assistant, playing in this way an instrumental role in the conception and production of Barbieri's publications *Artificial* (Edizioni Fotoselex) and *Silent Portraits* (Massimo Baldini Editore).

Claudio Alessandri has been an independent photographer in Vienna since 1988, he has won numerous prizes, works for ad agencies as well as on his own projects, such as the book *(women)**, Edition Stemmle, Zürich–New York, and has had various group and solo exhibitions.

"To me, photography is *the* chance to connect with people. Photography is an intimate moment of trust. Trust is the basis for freedom, and it is through freedom that human complexity emerges. If people trust me, I take a good photograph. There is something magical in this, and it is a privilege because in this way I am allowed to see the whole world.

"Photographing normal women and letting their beauty shine through is something that fascinates me. I am touched when these women see themselves, see how absolutely beautiful they are.

"My goal is quality and I want to leave behind proof that I was here and some proof of what I was capable of."

A-1030 Vienna, Fasangasse 49a/27
Tel. (+43-1) 799 12 97
Mobile (+43-664) 180 46 09
claudio@alessandri.at
www.alessandri.at

Mathias Bölinger, Interviewer

I was born and grew up in cozy Frankfurt am Main, after graduating from high school, I started to explore the then still untamed East. I opted against military service and performed my compulsory alternative public service in Moscow at a home for disabled children and with a mobile care unit for the elderly. Slavic studies at the University of Vienna. Later I added Sinology. After spending a year studying Chinese in Shanghai, I returned to cozy West Germany, where I continued my studies at the University of Cologne. At some point I wrapped up my studies, doing internships and volunteer work for various newspapers in between, including the *Frankfurter Rundschau*, the *Neue Zürcher Zeitung*, the Swiss culture journal *du*, and the *Deutsche Welle* radio station. Now I am an independent journalist with a weakness for eastern Europe and Asia and a strong interest in transformation processes and conflict regions.

D-50937 Cologne, Palanterstraße 5a
Tel. (+49-221) 484 81 21
mathias@boelinger.net
www.boelinger.net

Boris Bonev, Production in Austria

Born in Vienna in 1965. His mother picked his first name, though it is his father who comes from Bulgaria. Artistically inclined from early on, he spends a restless childhood and youth, and later trains to become a lithographer and graduates. Early on, he notices his inclination toward perfectionism, and through experiences and self-discovery, through an inquisitive nature that causes him to search for deeper meaning in all things, he develops a practical-philosophical way of life that also includes yoga. He enjoys working with people, creating things together, and watching them grow, and this has become a fundamental aspect of his life. All levels of his life are infused by a longing for perfection: from printing and all its preliminary stages to teaching yoga or managing an Austrian corporation to his private family life and the many ways he perceives day-to-day existence and the abundance and opportunities it has to offer.

Karl-Millöckergasse 1–9/12
A-2353 Guntramsdorf
Tel. (+43-664) 832 46 54
bonev.boris@aon.at

Hans-Joachim Gögl, Editor

Born on the banks of Lake Constance in 1968; trains to be a book sales-man with a brief interlude working at a gallery in Munich; afterwards, freelancer at the culture department of the ORF state radio station in Vorarlberg and occasionally for Ö1. Since 1992, independent public rela-tions consultant in Bregenz – with the good-natured and patient Monika Stelzl at his side – and independent father of Moritz and Marie Gögl.

Along with his public- and private-sector work, his most marked projects have been organizing "TRI," an international architecture sym-posium for energy-efficient building, which has taken place every two years in Bregenz since 1996; since 2003, publishing, designing, and co-hosting "Tage der Utopie," a lecture series that explores politico-social models for the future; and since 2004, together with Clemens Theobert Schedler, project director of the book and seminar series "Landscape of Knowledge." Was shortlisted for the Austrian State Prize for PR in 1999 and for marketing in 2004.

Sometimes I regret not being a specialist. One of those people who know a lot about a little, whose biography gives the satisfaction of stak-ing a claim, one's own professional territory where one knows every phenomenon, every inhabitant, every new finding down to the small-est detail. I take comfort in the thought that I am perhaps an expert in mediation, wandering back and forth like a nomad between thematically distinct fields of knowledge, equipped simply with the ability to become passionately enthusiastic about people and their expertise, to ask ques-tions and formulate the answers given by specialists in such a way that they have an impact on me and people like me.

A-6900 Bregenz, Belruptstraße 17
Tel. (+43-5574) 44729
hansjoachim@goegl.com
www.goegl.com

Eva Lepold, Research

Born in 1970 in Vienna, a town that always seems to draw her back again. Studied drama, Romance languages, linguistics, and English philology, dabbled in painting. Studied abroad in Manchester and Montpellier. Since 1996, has worked in public relations and international lifestyle with emphasis on event management.

From 2001, fled the city and went to Burgenland in pursuit of a vision of a life among sheep and plants. This led her to Butoh dance and freelance work in the publishing sector: numerous translations from English to German, particularly in the area of body and energy work, meditation, and philosophy of life interspersed with research, layout, and graphic jobs.

Since 2003, flees the countryside and goes back to Vienna. Training in play and experiential learning models, which brings her back to nature. In 2005 she was involved in a teaching project to help underprivileged children in the cloud forest in Ecuador. Currently she is involved in social work setting up a youth club and training to become a dance instructor.

A-1020 Vienna, Castellezgasse 14/15
Tel. (+43-1) 216 26 60 or
        (+43-699) 128 928 16
eva@metamorhpa.at

Kimi Lum, Translator

Translator since 1995. Main focus: film, art, architecture. I have always been intrigued by the written word but didn't discover my love for translation until I made Vienna my home base a decade and a half ago. Self-employed. Graduated from the Center for Translation Studies, University of Vienna. Came to Europe at the age of 20, grew up in California. Born in Honolulu, Hawaii, USA.

A-1030 Vienna, Ungargasse 65/9
Tel. (+43-1) 952 6332
kimi.lum@chello.at

Claudia Mazanek, Copy-editor in Vienna

Born in 1951 in Vienna. Passionate reader. Throughout her forays into various professions including physical therapy, studies in philosophy and political science, and academic project work, her gently enduring love for books has always and steadfastly remained, revealing to her in the end that the apparently lonely job of the copy-editor is a perpetually pleasurable and interlinking occupation. Book people of all kinds enrich our lives. And whenever the letters threaten to become overwhelming, *there is always a ship that awaits her, and the vast open sea...*

A-1080 Vienna, Albertgasse 41/4
Tel. (+43-1) 408 22 36
claudia.mazanek@gmx.at

Michael Prellberg, Interviewer

I never thought of Cassandra as an unhappy woman. How wonderful it must be to be omniscient! Who cares if no one believed her. What remains of those fantasies of knowing everything is my curiosity, the occupational disease of all journalists.

Unlike Cassandra I have to assume people might believe me, and so I must understand, and write what I have understood in an understandable way. The *Berliner Zeitung*, before that the *Harburger Nachrichten*, and now the *Financial Times Deutschland* have all published my work. I am grateful for that. Being a journalist is, at least to me, a dream job. A privilege.

For those who prefer hard facts: Michael Prellberg is an editor for the *Financial Times Deutschland* in Hamburg. He is married and has a fast-growing son, Hagen.

D-20357 Hamburg, Schwenckestraße 3
Tel. (+49-40) 319 90 – 242
prellberg.michael@ftd.de
www.ftd.de

Martin Rasper, Interviewer

Born in Brussels in 1961 as the son of a Tehran-born mother of Caucasus German descent and a Silesian father, grew up in Eppstein im Taunus, studied geology and philosophy in Munich and Berlin, worked as geologist, mailman, ad writer, copy editor for screenplays, book and newspaper editor. Since 2001 he has established himself as an independent journalist and writer in Munich and made a name for himself with topics and articles that oscillate between science and culture. He writes the monthly column "Wissenswelt" for the culture journal *du* and the interview column "Aussichten" for *natur + kosmos*; he also contributes regularly to *merian* and other magazines.

D-80638 Munich, Nymphenburger Straße 145
Tel. (+49-89) 18 17 74
rasperman@freenet.de

Clemens Theobert Schedler, Editor

I was born in Munich in 1962, grew up in Vorarlberg, have lived in Vienna since 1983, I am married to Márcia and am the father of three daughters, Lisa Katharina, Tereza Maria, and Elena Sofia.

At the age of 20 I started to learn the craft of visual design: one year of art history in Salzburg, four years at the Höhere Graphische Bundes- Lehr- und Versuchsanstalt in Vienna; three years of independent project work for Walter Bohatsch followed by more of the same for the ad agencies Demner & Merlicek and DDB Needham Heye & Partner. I went into business for myself in 1990 and was soon swamped by the storm and (literal) stress of the job and overwhelmed by the standards by which I sought to realize my ideas.

Burnt out at the age of 35, I went through an inner paradigm shift, sought *pull instead of pressure, plenitude instead of privation*, etc. The initial euphoria and the illusory assumption that from then on I would live comfortably and happily ever after gave way to a series of peaks and valleys and the realization that my happiness lay in the gloom of levity and in the brightness of death.

In my work, I simply try to bake good bread, to create resonance spaces for looking, seeing, reading, and feeling. To me, power lies in the dignity of the unpretentious, in the movement of the simple, and in the solution of the almost obvious. *A solution is something that is good for everyone.*

A-1140 Vienna, Leegasse 5/13
Tel. (+43-1) 440 50 60
Mobile (+43-676) 310 75 80
brief@konkretegestaltung.at
www.konkretegestaltung.at

Renata Schmidtkunz, Interviewer

My parents liked to tell about how, even when I was so small that I could barely talk, I always asked "why?" Some things never change, and this is one of those things. When I decided to study Evangelist theology at age 18, it wasn't because my father was also a priest, but because I was looking for my own reasons for the discrepancy between what so many people said about God and what I found to include the opposite of freedom.

I didn't plan to become a journalist, it was just the result of what had gone before, and it happened almost on its own. I have been working in television and radio for ORF (Austrian Broadcasting Company) for 17 years as an editor, filmmaker, and presenter. I started my career with ORF working for the Religion Fernsehen department (television), later I added Ö1 (radio) and 3sat (television), where I have also been working for many years.

My main fields of focus are culture and science. One program I passionately enjoy hosting is "Im Gespräch" on Ö1, which I started out presenting under Peter Huemer.

ORF 3-sat und Radio Österreich 1
A-1136 Vienna, Würzburggase 30
Tel. (+43-1) 87 878 13 359
Mobile (+43-664) 817 817 1
Renata.Schmidtkunz@orf.at

Samuel R. Schubert, Interviewer

American political scientist and author. His studies include academic journal pieces on economic and political development, international security and public policy. He served as an expert on many international economic development and cooperation programs for the United Nations between 1990 and 2000 before returning to academia as a researcher in the social sciences. He has since received fellowships in the United States and Europe and frequently lectures on contemporary political affairs. In the 1980s, while studying at George Washington University, he worked for the National Science Foundation, the country's renowned network of academic–industrial cooperation, in Washington DC, where he was directly involved within America's technical enterprise.

His most recent works include "Revisiting the Oil Curse" published in the September 2006 edition of *Development* (Palgrave Macmillan) and "The Asymmetry of Political Violence, Terrorism and the Terrorist" published in *Asymmetrie im 21. Jahrhundert. Reflexion über ein Zivilisationsphänomen im Kontext der Globalen Sicherheit* (Nomos, 2005). An updated version of that piece was published by Simon and Schuster in New York in 2007. He is frequently commissioned to author strategic country forecasts for the insurance and risk-assessment industry and is currently working on two larger projects, a history of the modern Middle East and the international security implications of demographic changes in Europe.

A-1020 Vienna, Karmeliterplatz 1/21
samuelschubert30@webster.edu

Landschaft des Wissens:
Association for the Promotion of Science, Business Culture, and Regional Development

Landschaft des Wissens was founded in 2004 in Klagenfurt, Austria. The members of the association are active figures in Carinthia's economic life who are concerned with the theory and practice of development strategies for rural areas. The association is not affiliated with any political parties, its projects are funded by grants from the European Union, the KWF – Kärntner Wirtschaftsförderungs Fonds, and sponsors.

Its main objective is the researching, discussion, initiation, and guidance of innovative projects for cooperation. The concentration on a culture of cooperation is based on a specific historical experience and on the assessment of a current and future necessity: Carinthia, a region whose cultural diversity is typical for Europe, now stands at the junction between the German-, Italian-, and Slovenian-speaking worlds – formerly with its back to the Iron Curtain; today with front-row seats at the central and eastern European stage of EU enlargement. In this sense, cooperation is a skill we grew up with and one that is alive in the minds of a large part of the population, but at the same time and especially in these border areas there is a deep-seated culture of fear of the proximity of the "foreign" as well as some very real potentials for conflict.

The economic situation of Austria's southernmost federal state is marked by the absence of strong urban centers – parallel to a small number of large industrial enterprises there is a successful decentralized structure of many small and medium-sized businesses in the gastronomy, trades, and commercial sectors as well as a growing structure of IT companies (see the section on Lakeside Science & Technology Park in Klagenfurt, pages 127–154).

With increasing rationalization and centralization in a global market, the ability of regions, enterprises, and public institutions such as universities to cooperate with one another has become an important competitive factor.

Network management, clustering, temporary consortia, industry associations, research and development cooperation projects, etc. – the professional management of the cooperation between independent small units is by no means to be regarded as just a defensive reaction to the efficiency of industrial production or centralized growth.

Development in information and communication technology or in the ability to run complex structures through increased managerial training is already bringing forth organizational models that are on a par with the interaction between the departments of a corporation or the institutions of a metropolis. The benefits of consolidation – in its urban or economic form – are being implemented increasingly in rural areas or between small enterprises. These are models that not only play an important role in the interviews presented here and the projects covered in the first volume, *Big Strategies for Small Business*, but which will also serve as subjects for discussion to everyone involved in the field of regional development.

The concrete activities of the association

These are threefold – publication, discourse, and project development:

The publication series "Landschaft des Wissens" investigates cases which from our perspective constitute exceptional European projects on economic strategies for rural areas. Thus, on the one hand, we are collecting experiences on the development of rural areas from very different regions throughout Europe and bringing them to Carinthia, and on the other hand, we are sharing these experiences with other regions through this publication series. Each book will be published as separate unabridged issues in German and English.

The symposium "Landschaft des Wissens" is a biennial conference that will address the theory and practice of regional economic development, provide the opportunity to share experiences, and promote networking among the participants. The first international symposium took place in the spring of 2006 at Lake Weißensee. Entitled "Das Handwerk der Zukunft" [Crafts of the Future], this three-day conference brought together master craftspeople, small-business entrepreneurs, regional developers, and business consultants from the entire German-speaking world, who came to discuss new models for the sustainable development of their regions and operations. The symposium spawned a series of initiatives and ideas, but, above all, it forged relationships between participants, and many of these ties continue to make an impact today. The organizers are currently planning a symposium on the subject matter of the book you have before you.

The workshop "Landschaft des Wissens" will focus on the insights gleaned and their practical application within the regional context of Carinthia. Its aim is to introduce the heads of industry and academia to selected strategies provided by actual researched examples and to initiate project development or guide participants through the development process. Thus, each book serves as an impetus for regional research and development, while the workshop provides a concrete implementation structure.

Landschaft des Wissens
Association for the Promotion of Science,
Business Culture, and Regional Development
Lakeside B01
A-9020 Klagenfurt
Tel. (+43-463) 22 88 22-0
office@landschaft-des-wissens.org
www.landschaft-des-wissens.org